Oxford Chemistry Series

General Editors

P. W. ATKINS   J. S. E. HOLKER   A. K. HOLLIDAY

Oxford Chemistry Series

1972
1. K. A. McLauchlan: *Magnetic resonance*
2. J. Robbins: *Ions in solution (2): an introduction to electrochemistry*
3. R. J. Puddephatt: *The periodic table of the elements*
4. R. A. Jackson: *Mechanism: an introduction to the study of organic reactions*

1973
5. D. Whittaker: *Stereochemistry and mechanism*
6. G. Hughes: *Radiation chemistry*
7. G. Pass: *Ions in solution (3): inorganic properties*
8. E. B. Smith: *Basic chemical thermodynamics*
9. C. A. Coulson: *The shape and structure of molecules*
10. J. Wormald: *Diffraction methods*
11. J. Shorter: *Correlation analysis in organic chemistry: an introduction to linear free-energy relationships*
12. E. S. Stern (ed): *The chemist in industry (1): fine chemicals for polymers*
13. A. Earnshaw and T. J. Harrington: *The chemistry of the transition elements*

1974
14. W. J. Albery: *Electrode kinetics*
15. W. S. Fyfe: *Geochemistry*
16. E. S. Stern (ed): *The chemist in industry (2): human health and plant protection*
17. G. C. Bond: *Heterogeneous catalysis: principles and applications*
18. R. P. H. Gasser and W. G. Richards: *Entropy and energy levels*
19. D. J. Spedding: *Air pollution*
21. P. W. Atkins: *Quanta: a handbook of concepts*

E. S. STERN (Editor)
COMPANY PLANNING DEPARTMENT, ICI

The chemist in industry (2):
human health and plant protection

J. F. Cavalla
D. Price Jones

Clarendon Press · Oxford · 1974

Oxford University Press, Ely House, London W.1

GLASGOW NEW YORK TORONTO MELBOURNE WELLINGTON
CAPE TOWN IBADAN NAIROBI DAR ES SALAAM LUSAKA ADDIS ABABA
DELHI BOMBAY CALCUTTA MADRAS KARACHI LAHORE DACCA
KUALA LUMPUR SINGAPORE HONG KONG TOKYO

PAPERBACK ISBN 0 19 8554168
CASEBOUND ISBN 0 19 8554540

© OXFORD UNIVERSITY PRESS 1974

All rights reserved. No part of this publication may be reproduced, stored in a retrieval system, or transmitted, in any form or by any means, electronic, mechanical, photocopying, recording or otherwise, without the prior permission of Oxford University Press

PRINTED IN GREAT BRITAIN BY
J. W. ARROWSMITH LTD., BRISTOL, ENGLAND

Editor's foreword

Human health and plant protection is the second of the books in the Oxford Chemistry Series devoted to chemistry in industry. Their aim is to provide the necessary background information to enable the graduate entering the chemical industry to have an appreciation of current industrial developments and philosophy. This latest addition to the series will interest all those students who are considering entering the pharmaceuticals or agrochemicals industries or who are concerned with biological applications of organic chemistry which are making such a vital contribution to the evolution of modern society.

In recent years modern methods of synthetic organic chemistry have allowed organic structures to be manipulated and altered in directions that were not possible hitherto. The biological effects industries are being increasingly influenced by this situation and investigations of the relationships between structure and biological activity are made much easier by the availability of a wide range of 'tailor-made' compounds. In this book the authors develop the fundamental philosophy underlying the evolution of organic compounds with desirable biological effects. It is this philosophy which the synthetic organic chemist has to master before he can hope to make a real contribution in the biological effects industries.

The editor of 'the chemist in industry' group of books, Dr. E. S. Stern, has been closely associated with many areas of development in the modern chemical industry. His experience ensures that each of the team of authors is an expert in the subject of his contribution, so that the reader benefits directly from the authors' experience.

J.S.E.H.

Preface

ORGANIC chemistry has changed society. Elaboration of complex molecules by synthesis has provided 'effects chemicals' of remarkable sophistication. New colouring agents, brighter and more 'cost-effective', and other auxiliaries affecting the stability of materials, have been found, manufactured, and sold in the last fifty years. They make life easier and more pleasant. Chemicals with biological effects have changed man's attitude to hunger and illness within a remarkably short period. The industries concerned with pharmaceuticals and 'other biologicals' appeal to many students of chemistry who feel the call to serve society by means of their science. The effects they look for include the cure of ill-health, the promotion of health, the avoidance of overpopulation caused by undesired pregnancy, and increase in food production by pest control: much has been achieved, and there are a good many targets still to be reached, obvious to the discerning.

How does the industry find new products, and what is the role of the chemist in the industry? These questions, following naturally from an initial interest, are answered here from within industry. This book is meant to help students understand the industrial process and some constraints upon it, and to let them see what has been or can be accomplished and why other things cannot be tackled in industry today. We have two eminent contributors to this book: Dr. Cavalla, Director of Research and Development at John Wyeth and Brother, is a former Chairman of the Fine Chemicals Group of the Society of Chemical Industry and Dr. Price Jones, a writer on biological aspects of the agrochemical industry, is a former President of the Association of Applied Biologists. They have selected interesting and important topics for discussion; they paint the industry in its true colours and with proper perspective. We hope our purpose in informing and helping the reader is achieved.

E. S. STERN

Contents

CONTRIBUTORS	viii
1. BACKGROUND	1
2. THE PHARMACEUTICALS INDUSTRY TODAY	7

Introduction. The design of drugs. Field of science approach—biology. Field of science approach—physical chemistry.

3. SPECIFIC BIOLOGICAL EFFECTS	19

The penicillin antibiotics. Sulphonamides. Hormones—chemicals found in the body. Minor tranquillizers. Major tranquillizers and antidepressants. Antihypertensives.

4. THE INDUSTRIAL DEVELOPMENT OF A NEW DRUG	36

Team-work. Screening of chemicals. Safety of drugs. Clinical trial. Formulation. Marketing.

5. THE FUTURE OF THE INDUSTRY	42
6. THE AGROCHEMICALS INDUSTRY TODAY	45

Introduction. Research on agrochemicals.

7. SPECIFIC BIOLOGICAL EFFECTS	50

Control of pests—general. Control of insect pests. Control of Rodents. Control of plant diseases. Control of weeds. Regulation of plant growth.

8. NUTRITION OF CROPS	66

Nitrogenous fertilizers. Phosphate fertilizers. Potash fertilizers. Compound fertilizers. Liquid fertilizers.

9. INDUSTRIAL DEVELOPMENT OF A NEW AGROCHEMICAL	71

Pesticides and plant growth regulants. Fertilizers.

10. AGROCHEMICALS AND THE ENVIRONMENT	77
11. THE FUTURE OF THE INDUSTRY	81

Pesticides. Plant growth regulants. Fertilizers.

FURTHER READING	84
INDEX	85

Contributors

CHAPTER 1
E. S. Stern, *Imperial Chemical House, Millbank, London.*

CHAPTERS 2–5
J. F. Cavalla, *Research Director, John Wyeth and Brother Ltd, Taplow, Maidenhead.*

CHAPTERS 6–11
D. Price Jones, *Consultant Biologist, Reading.*

1. Background

E. S. STERN

THE industry discussed in this book is concerned with producing biological effects by suitably formulated chemical agents. It aims to prevent and cure disease in certain organisms on the one hand and to destroy pests and pathogens on the other. In principle, there is nothing new about this activity. Both medicinal and pest-regulating chemicals have been known for centuries: their origins are lost in antiquity. But over the last eighty years or so advances in science and, later, in technology, have led to a complete restructuring of the industry with profound effects on society.

Two closely similar but distinct industries can be recognized. The pharmaceutical industry deals with human health and has a direct bearing on the welfare of pets and even farm animals. The agrochemical industry deals with the regulation of pests in the widest sense; it also has a direct bearing on farming practice, animal health and nutrition, and improvement of crop growing. Both the pharmaceutical industry and the agrochemical industry resemble the aerospace and petrochemicals industries in their high content of advanced technology. They have a rapid effect on the well-being of society. Table 1 gives an indication of the size and growth of the pharmaceutical industry over a decade.

TABLE 1
Growth of the pharmaceutical industry

	U.K.		U.S.A.	
	Gross output ($m)	Manpower ('000)	Product ($m)	Manpower ('000)
1958	450	60	3000	104
1963	625	69	3700	112
1968	850	73	5300	125

The biological effects industries differ from aerospace and petrochemicals industries, however, in important ways. Thus, except for the manufacture of fertilizers they are not 'capital intensive': manufacturing plant producing these chemicals forms only a small component of the total investment in the industry. Much of the investment is in accumulated research and development expertise, in the accumulated knowledge of the needs of the potential customers, and in marketing skills, i.e. in getting the products to meet the needs of the customers. Compared with the sales value of its product, the fixed investment in the industry is not large. In many cases the sales value per ton of active chemical ingredient is very high, partly because the ingredients are very active, and partly because the effects are very desirable. Especially if his own health is at risk, the customer is prepared to pay a high price. Who can put a value on the cure of Sir Winston Churchill's pneumonia during the Second World War? Indeed pneumonia provides a good example of the effect of the industry on society: the number of deaths from pneumonia in the U.K. in people under 65 dropped from about 22 000 in 1930 to about 6000 in 1960, almost wholly because of drug therapy. Over 300 000 people are now alive in the U.K. who would have died of pneumonia without chemotherapy. Such a saving has a major effect on the total population, and on the national economy.

How were new chemical treatments found? Even in early times there were highly effective biological agents, though their mode of action was not understood. Many an old wives' tale has been shown to be soundly based—eating liver against anaemia; using moulds in cuts and grazes; using certain natural extracts against fevers, particularly cinchona bark extract against malaria; and using pyrethrum extracts against insects. The modern biological effects industry has used these and other leads to provide effects that were predictable and selective, that cured with certainty and did not kill. The road to this point of achievement has been difficult and setbacks have occurred, particularly in isolated cases when unsuspected damage has belatedly been recognized. No one, however, can deny the progress made and the achievement of the industry.

The biological effects industry is based on research done towards certain ends or objectives. The objectives of the research are to meet human needs and they change as old needs are met and new needs are recognized. Forty years ago the need was overwhelmingly for chemotherapeutic agents to combat bacterial infections. A wide range of chemicals is now so effective against bacteria that further research along this line is likely to offer diminishing rewards. As a result, research has tended to move towards other needs or markets. Often the successful conclusion of a research line throws up a new market opportunity: for example, people who might have died from bacterial infection now survive to an age where their blood circulatory system deteriorates and they may require remedial drugs.

Thus, although the industry is research-based, it is very much market-oriented. The effort required for success, i.e. the cost of producing a new agent, is very large and statistically predictable: if the market is too small to permit recovery of this cost, then commercial research and development (R & D) directed at the target is ill-founded. An R & D project to cure elephants of an obscure disease is unlikely to be an economic success: too few people will wish to pay for the cure. Prevention of fungus attack on barley offers a target that can be objectively assessed—the tonnage lost can be estimated and provides the economic basis for the programme. Certainly the grower will not spend as much on the remedy as this potential loss: he hopes to gain by the innovation. The effect has to be demonstrated and the gain spelt out. Human health offers a different problem. The customer of the industry is usually the doctor who prescribes the treatment, not the patient who takes the cure: to him the cost is less important than the efficacy and safety of the treatment both of which have to be demonstrated convincingly. Today, it is the expert in the market (often comprising a committee set up by Government) who judges the acceptability of the product and the industry must anticipate his every question and problem.

The question 'Who is the customer?' is a vital one to answer, if waste is to be avoided. It is particularly difficult in the pharmaceutical industry—in fact the industry is split in two. One part, often called the 'ethical' pharmaceutical industry, is concerned with the medical profession: its products are prescribed for the patient by the doctor. The products include drugs curing disease, medicines ameliorating disorders, and 'preventatives' that keep the body healthy.

The second part of the pharmaceutical industry provides the so-called 'proprietaries'. These are bought by the ultimate user from pharmacists or drug stores. Some are folk medicines, some are ethical drugs suitably formulated for the layman so that mild misuse causes no damage. Clearly poisonous substances or dangerous drugs cannot be sold in this way. The importance of the distinction between these two parts of the pharmaceutical industry lies in the different selling techniques. Highly qualified people are needed to demonstrate new advances to the medical profession, but the man in the street will buy a proprietary medicine on the basis of a television or newspaper advertisement appealing to him, or on a pharmacist's recommendation.

The pest-regulating industry also sells to widely dispersed consumers, often through intermediate co-operative purchasing agencies. Its main outlet is the farmer, a very critical cost-conscious customer. From the marketing point of view, there is a reasonably close resemblance between the two: both try to achieve predetermined biological effects chemically and require a major scientific testing programme for the accurate evaluation of this effect and any accompanying undesirable side-effects. In both industries, increasing attention is being paid to side-effects. Extensive biological testing is essential to make certain whether treatment is ultimately desirable or harmful. These tests greatly add to the cost of the product as finally marketed. The expenses incurred are regarded as part of the R & D expenditure of the firm and Table 2 shows how different the make-up of the product cost is in the pharmaceutical industry from that in the petrochemical industry.

This high R & D expenditure, including the high cost of ensuring the safety of the material, makes it very difficult to get onto the market the new products on which the industry has so greatly depended for growth over the

Table 2

Cost breakdown of product—in percentage terms

	Pharmaceutical industry*	Petrochemical industry
Manufacturing cost	40	60
Marketing costs	20	8–10
R & D expenses	10	~4
Admin. and similar	10	~10–12
Profit	20	15

* Sainsbury Report on the Relationship of the UK Pharmaceutical Industry with the UK National Health Service (HMSO, London 1967).

Table 3

Relationship of single chemical entities to total new drugs introduced in the United States between 1959 and 1968

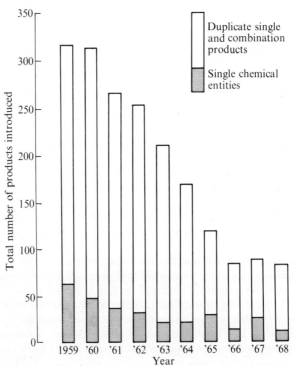

TABLE 4

Composition of the ethical pharmaceutical industry in 1971

	U.S.A.	Japan	Germany	France	Italy	Spain	U.K.	Belgium	Holland
Total market (retail outlets) $ million	3640	2456	1202	1167	1017	495	437	228	135
Number of Companies selling									
$30+ million	36	26	9	3	2	—	2	—	—
$10–30 million	34	31	23	30	25	5	12	1	1
$5–10 million	12	16	25	38	26	22	18	10	6
$2·5–5 million	16	20	41	35	39	34	9	15	7
$1·0–2·5 million	15	59	137	32	35	84	141	145	168
Per cent of market covered by listed companies	98·4	99·5	98·4	96·9	93·9	93·9	100·0	97·9	95·6
Top company $m	278	154	46	47	42	14	41	11	12

last few decades. In addition, the recent emergence of regulatory authorities empowered by law to permit or forbid entry to the market of new medicines has dramatically diminished new product introduction. Table 3 shows the number of new drugs introduced in the U.S.A. in the period 1959–1968 and how this number is dropping. The pharmaceutical industry is thus caught between the rising cost of ensuring the safety of a drug and the diminishing chance of producing a new drug: the amount of R & D per successful drug is rising steeply. Indeed, it has risen in the last decade from about 7–10 per cent of total sales to about 15–20 per cent. As a result only the largest firms can afford sufficient R & D to sustain a regular flow of innovation. The biological effects industry is thus made up of a moderate number of very large firms relying on innovation and of a much larger, but diminishing, number of smaller firms purveying more traditional products. Table 4 illustrates this for nine major countries showing how, with the growth of the size of the market, productive capacity is vested evermore in the larger units.

This introduction should place the two industries in perspective and we can now turn to a more detailed consideration of the industries themselves and of the role of the chemist in them.

2. The pharmaceutical industry today
J. F. CAVALLA

Introduction

THE human health industry, like all other industries, is a product of its past. Its earliest members were successful pharmacists who had found themselves incapable of satisfying the demands for their nostrums and extracts by simply working harder in the back of their shops. First, small factories started, then reputations were made, such that the Jesse Boots, Henry Wellcomes, and David Hanburys of this and other countries became household names; as a result the industry prospered and grew. It was based upon the ability to extract natural products from mainly vegetable sources and to compound them with sufficient reliability in a variety of forms for a guaranteed effect to be produced. To this work was added, in the early part of the twentieth century, the preparation of serums and antitoxins.

Gradually, and with the discovery of sulphonamides more rapidly, the industry changed from extraction to synthesis, from domination by the pharmacist to domination by the chemist. Some of the newer firms derived from offshoots of the mammoth chemical companies which had been brought to prominence by the advances of organic chemistry in the nineteenth century: the Dyestuffs Division of I.C.I. in Britain, that part of I.G. Farbenindustrie which is now Hoechst in Germany, and Rhône–Poulenc in France.

Later still, during and after the Second World War, other entrants appeared with the discovery of penicillin, streptomycin and tetracyclin. Companies skilled in fermentation technology, for example of citric acid, diversified promptly into the fermentation of natural mould antibiotics.

Except with hindsight, no natural progression can be seen in the establishment of these pharmaceutical firms. Their growth was based on successful innovation. Thrift, hard work, and efficiency are no substitutes for inventive capacity. The target—improved health—was always there, what was needed was the achievement of the effect. Without new products a company cannot grow in the pharmaceutical industry and will slowly perish, while occasionally, with one dramatic invention, a minnow in the industrial shallows can develop into a whale in a dozen years. As a result and as a determinant for survival the industry has become research-centred and, predictably, will so continue. The whole industry now reflects this: more graduates are employed per unit of production in the pharmaceutical than in any other industry.

It has been said that if you make a mousetrap better than anyone else, though you hide in the depth of the forest the world will beat a path to your door. This may have been true in the nineteenth century when Emerson first proposed it, but it is not the case for the pharmaceutical industry today. Although success must depend on having an effective product, it is equally important to establish it effectively in the market—and in competition with other companies' products with similar activity.

As the industry concentrates on innovation, research costs rise. Extrapolation of the present trend suggests that by the turn of this century the industry world-wide will be dominated by twenty or so international companies which alone will be able to break new ground. With the growing complexity of the therapeutic problems remaining, the big groups should have dramatic advantages over the small. Not the least of these is the capacity to exploit an invention once it is made. The patent, guaranteeing sole exploitation rights, runs less than twenty years in most countries. At least half this period can be taken up in getting the compound to market. So only a relatively short time remains for commercial reward to be gained, indeed, for recovering the development expenditure. A world-wide 'multi-national' company has access to a world-wide market and the large cash flow needed to establish the innovation quickly. Smaller companies have difficulties with timing, licences, contractors, agents, and the control of non-indigenous sales forces.

The smaller companies will not all necessarily perish. There will always be imitative research and a sale for useful drugs when they become free of patent embargoes; and there will always be fringe medicines, surgical requisites, infant foods, and slimming aids permitting non-innovational concerns to survive. But our concern is with the innovational research-based companies that have made the pharmaceutical industry what it is today.

The design of drugs

In designing a drug the chemist has as his main objective the cure of a specific disease by chemical means. A secondary, yet most important, objective remains: to improve those cures we already have. Such improvement can either be technical i.e. more activity and fewer side effects, or commercial i.e. to provide a product as good at a lower price.

There are three main pathways the chemist can follow in making a substance which he hopes will turn out to be a drug. First, he can synthesize a variety of arbitrary structures and, by serendipity, find that one or other of them is active. For success he will need a large measure of luck and some very able biologists, especially pharmacologists, to set up animal models resembling human adverse conditions so that the effect becomes obvious. The second, slightly more effective, method is to start with a compound which already has some effect, however marginal, in the animal model; then, by the application of skill and experience, so modify that structure that activity and potency are enhanced and toxic side-effects diminished. In this case the chemist uses the animal test to optimise the structure–activity relations empirically. The third method entails team-work between biochemists and biologists of various types to examine the fundamental nature of a body process or malfunction on the assumption that a chemical can be designed to affect the former or correct the latter. This is long-term research which, if successful, offers the opportunity of a major advance. It requires the co-ordination of a firm's total resources—scientific, commercial, and financial.

Examples can be given of successes achieved using each of these methods. With the increase in knowledge needed to tackle more and more difficult problems, it is the third method which has, in recent years, been receiving increasing emphasis, especially in the larger firms able to marshal the necessary resources towards a major advance, say in circulatory diseases, disturbance of the central nervous system, or hormonal biochemistry.

The chance approach

This heading encompasses all the folk remedies and natural products discovered throughout man's history. Far from being of little value, many of these continue to be used. Quinine was used as a crude extract (of cinchona bark) long before its structure **1** was known; so also morphine **2**, atropine **3**, and reserpine **4**. For each of these, wholly synthetic alternatives now exist

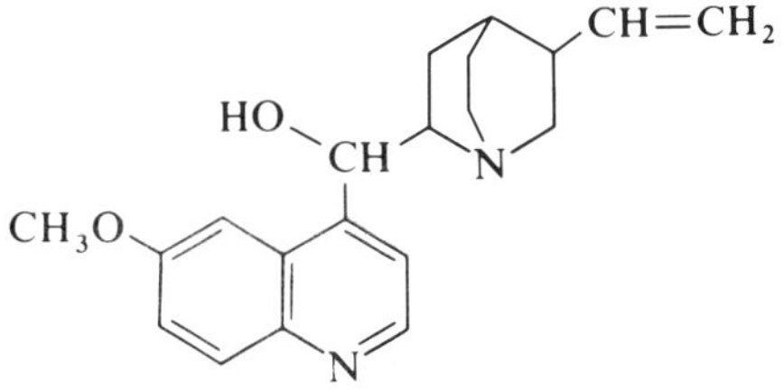

1 Quinine

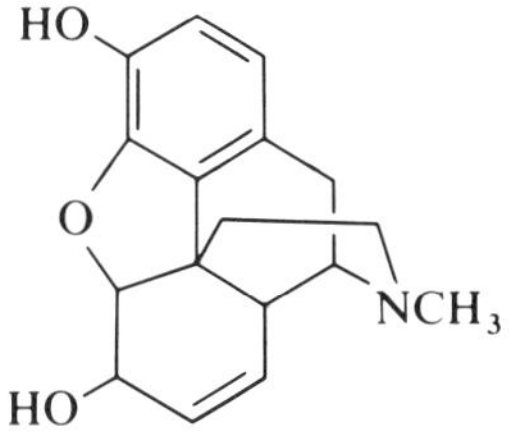

2 Morphine

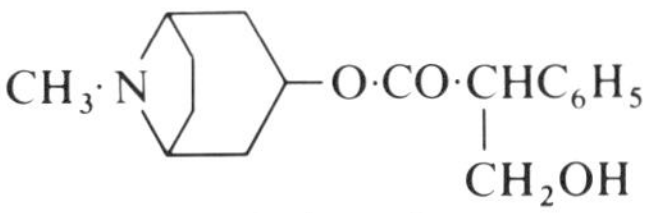

3 Atropine

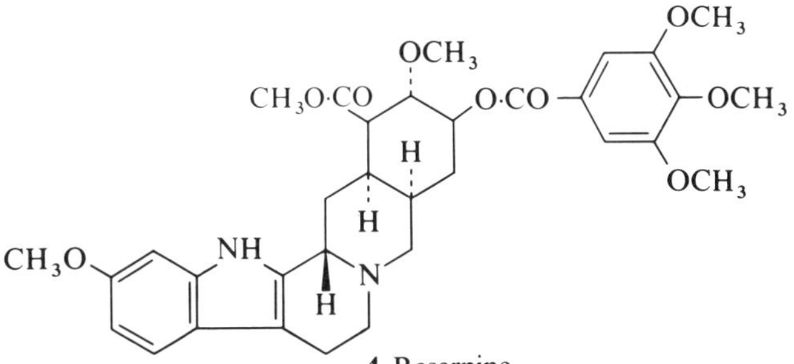

4 Reserpine

which are claimed to be superior. But in spite of this, the first two, quinine and morphine, remain pre-eminent in the cure of malaria and the relief of pain, respectively.

Many examples can be given of synthetics, made either speculatively or for some other purpose by the chemist, which owe their success in medicine to the skill of the pharmacologist in noting an unexpected effect in animals. One early example was the pain-killer pethidine **5** made by Eisleb in 1939 as

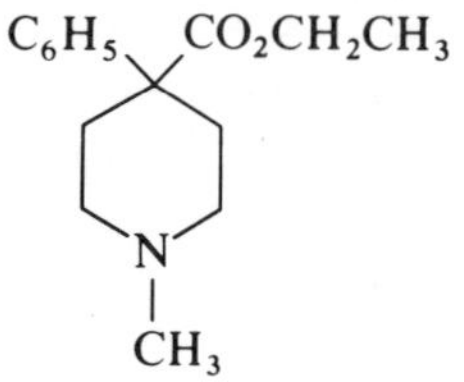

5 Pethidine

an analogue of atropine and recommended by him for testing as an anticholinergic agent. It was the pharmacologist, Schaumann, who observed the analgesic property of the compound and it is as an analgesic that it is now widely used. The anticholinergic effect is a weak and essentially unwanted side-effect.

The complex structure of chlordiazepoxide **6** was made almost accidentally by the Hoffman–La Roche chemist, Sternbach. The pharmacologists of that company found that it had the tranquillizer properties which have rendered it so successful. Many hundreds of related benzodiazepines have been prepared since, several much superior to the model; but their discovery rests upon the lead provided by the unexpected observations on **6**.

Whenever chance provides a major discovery, scientists afterwards try to explain the logical basis of the work. There can be no doubt, though, that such widely sold drugs as phenylbutazone **7** for rheumatism, phenobarbitone **8** for epilepsy, and chlorpromazine **9** for schizophrenia originated following chance pharmacological or clinical observation: in no case was the compound made for the purpose for which it is now used.

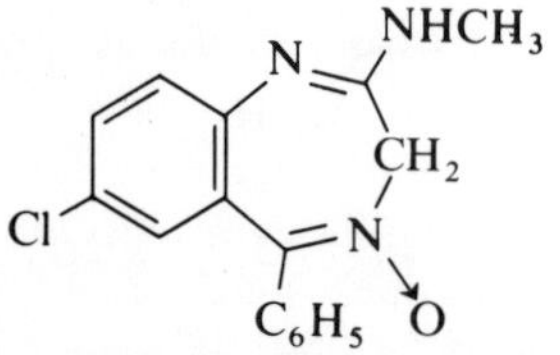

6 Chlordiazepoxide

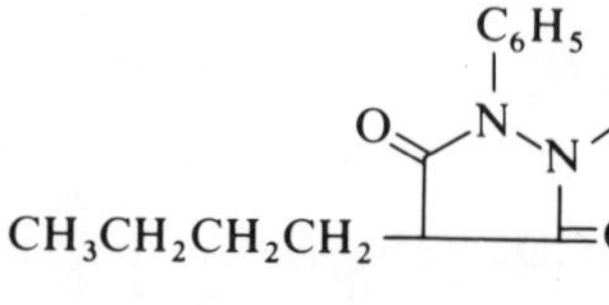

7 Phenylbutazone

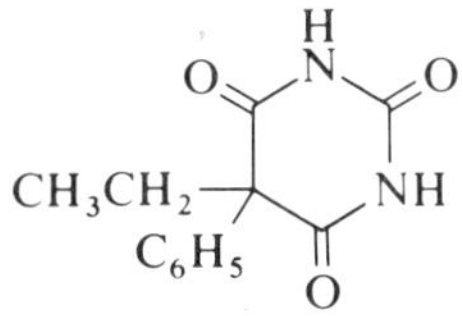

8 Phenobarbitone

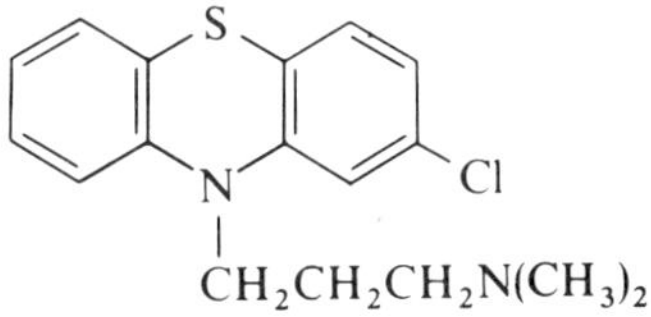

9 Chlorpromazine

Molecular modification of known agents

Many useful, even invaluable, drugs have been discovered by structural modification of known drugs. Such research is regarded by some critics as being below the attention of the 'true' scientist. Pejorative descriptions such as 'molecular manipulation' or 'molecular roulette' are applied without recognition that it is by the application of scientific method and by correlation of knowledge that leads are provided. It is true that trivial modifications are sometimes made to structures purely for commercial reasons to permit entry into a market dominated by a product patent; but if the new is not superior to the old it will fail. No amount of advertisement will allow the inferior product to succeed with customers as value conscious and as capable of evaluation as qualified physicians.

In most cases the objective in modifying a structure is technical improvement. Mention has been made of the natural product morphine, (**2**, p. 9), still widely used as a strong analgesic (as opposed to the weak analgesics such as aspirin and paracetamol). Yet it has well-known drawbacks, which could profitably be lost: its capacity to induce habituation, and the respiratory depression it causes, especially in the elderly. It also causes constipation, nausea, and sedation. But it was predominantly its addiction component that first attracted the medicinal chemist. Following the early recognition of the problem of addiction, the U.S. government requested from the National Institutes of Health at Bethesda a non-addictive analgesic in 1930. A group of chemists and pharmacologists was formed to tackle the task and has since been joined by like groups in most of the major pharmaceutical companies.

At the outset the objective was clear: a non-addictive analgesic. The approach chosen was so to modify the morphine structure that analgesic action was retained but addictive potential was lost. As well as the chemical problem, there was the problem of measuring quantitatively the biological effects. To test for the analgesic activity was relatively simple since small animals show up the action of morphine against experimentally induced pain. To show lack of addictive liability, however, was much more difficult: it was almost impossible to make a small animal dependent upon morphine. As an illustration of the difficulty it is salutary to recall that heroin **10**, the supremely addictive diacetate of morphine, was first introduced as 'non-addicting'.

The chemical approach to the problem was to simplify the morphine structure which had been proposed in 1926 by Robinson (proved only in 1952 by total synthesis by Gates). Simplification of the morphine structure was not difficult and soon structures were appearing based first on the morphinan skeleton **a** and then on that of benzomorphan **b**. Many of these analogues were many times more active than morphine itself but none proved superior: all were found to be addictive in man. There is an important lesson here for medicinal chemists: it is not the absolute potency or dose that

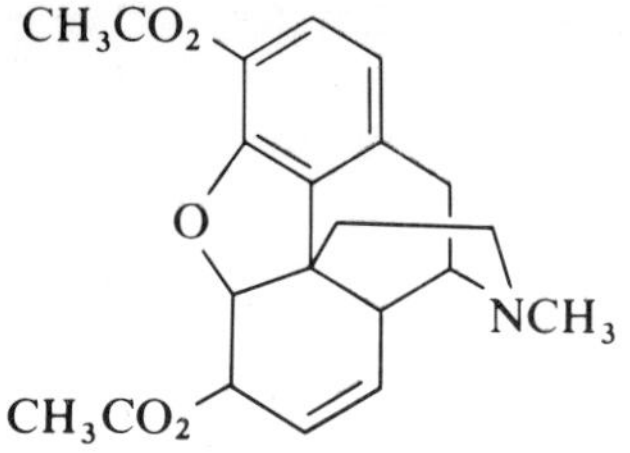

10 Heroin 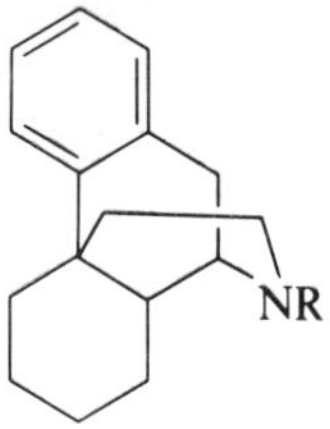   **a** Morphinan 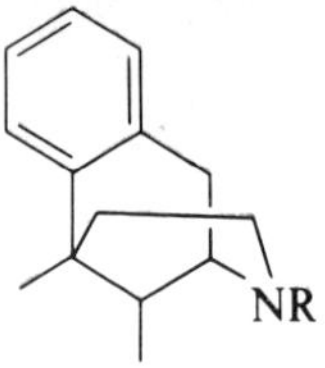   **b** Benzomorphan

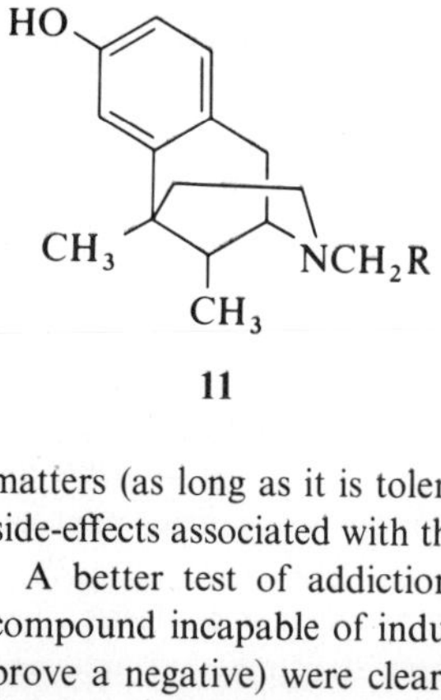

11 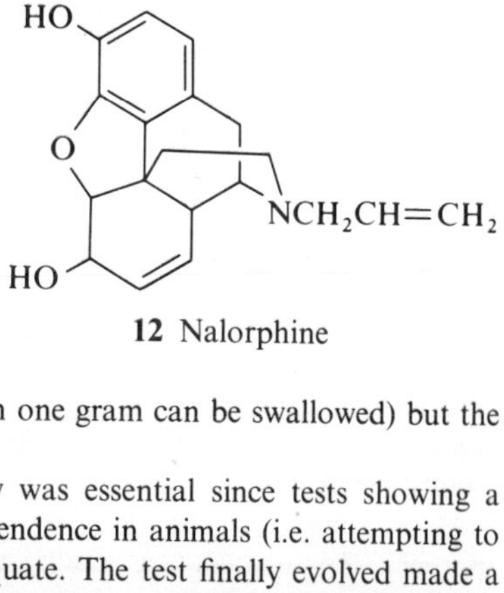   **12** Nalorphine

matters (as long as it is tolerable even one gram can be swallowed) but the side-effects associated with that dose.

A better test of addiction liability was essential since tests showing a compound incapable of inducing dependence in animals (i.e. attempting to prove a negative) were clearly inadequate. The test finally evolved made a monkey dependent on morphine and then determined whether the experimental compound was capable of substituting for morphine in maintaining the addiction. This more reliable test led to progress. It was found in 1958, that phenazocine (**11**, R = benzyl) though ten times as potent an analgesic as morphine in the mouse, was only one-fifth as effective as morphine in maintaining addiction in the monkey. Thus the compound appeared fifty-fold the superior of morphine. Unfortunately, when tested in man,

The pharmaceutical industry today 13

phenazocine was found to be little more analgesic than morphine and equally addictive. Such species-differences are often encountered. It appeared in this case that a compound to be safe would have to be completely devoid of the capacity to maintain dependence in monkeys.

The solution, when it came, derived not from science but from chance. An early analogue of morphine was nalorphine 12 where the N-methyl group had been replaced by N-allyl. This compound was inactive in mice as an analgesic. In fact, it had the capacity to antagonise an analgesia present and induced by morphine. Though it itself caused respiratory depression, it could antagonise the respiratory depression caused by morphine and it was used clinically for this purpose in cases of morphine overdosage. When given to addicts it precipitated the so-called 'withdrawal syndrome', i.e. it destroyed the action of morphine, sometimes with startling effects. On these grounds nalorphine was regarded as an antianalgesic, a morphine antagonist. With this in mind, two American clinical pharmacologists, Beecher and Lasagna, conceived the idea of dosing patients with a mixture of morphine and nalorphine. They hoped that by subtly modifying the dosage they could determine how much of the nalorphine would antagonize the undesirable respiratory depression and addictive liability of morphine, without seriously impairing the latter's analgesic capacity. In the event they found that far from impairing the analgesia, nalorphine actually enhanced it and from that observation it was a short step to the discovery that nalorphine, though essentially devoid of analgesic activity in the mouse, rat, dog, and monkey was an analgesic in man with a potency equal to that of morphine. Ostensibly the problem then appeared solved: nalorphine was an analgesic incapable of producing addiction. Though we now know it is possible to induce habituation with nalorphine in man, the habituation so lacks sensory satisfaction that it is immediately discharged and not renewed when drug dosage is stopped. Unfortunately, nalorphine proved to have a quite unexpected drawback: a hallucinatory effect was felt by many patients when given nalorphine to relieve their pain. So severe was this feeling of craziness that most people in whom it occurred declared they would rather suffer their pain than take nalorphine again.

However, the discovery of narcotic antagonism by nalorphine changed the character of the problem: the old approach was abandoned and it became desirable to find an antagonist devoid of hallucinatory properties. The antagonist activity seemed to be vested in the allyl group and allyl groups were therefore placed on all known analgesics. Most of those with morphine-like structures proved to be antagonists; others (pethidine for example) did not. Unfortunately, most also proved hallucinatory although compounds are now available which are claimed to be devoid of this effect. One of these, pentazocine (**11**, $R = CH:CMe_2$) has reached the market: it contains not a simple allyl group but a dimethylallyl. It is not as potent as morphine as an analgesic nor as potent as nalorphine as an antagonist, but as the first non-addicting strong analgesic offered it has achieved an established position. Two derivatives of 14-hydroxynormorphinone(**13**, R = CH:CH$_2$; and **13**, R = cyclopropyl) are also potent antagonists but have little analgesic power and in this regard may prove to be antidotes of some promise.

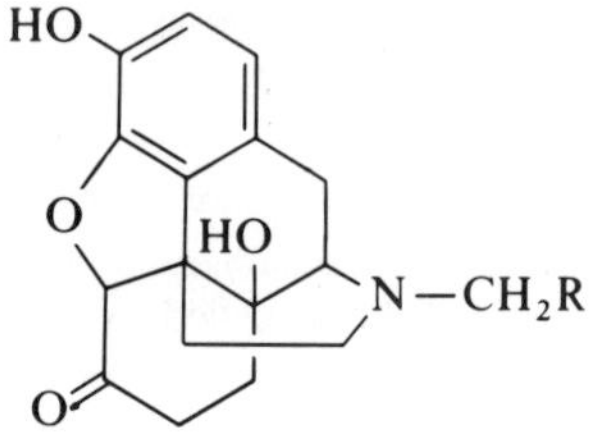

13

Field of science approach—Biology

In view of the spectacular 'instant' successes achieved by chance observation and empirical correlations, examples of drugs based on long-term research seem almost pedestrian. Yet it is this approach which progressively is occupying most research teams in the pharmaceutical industry. In some cases the history of the subject goes back many years and here must be greatly compressed.

Drugs based on histamine. In 1910, Sir Henry Dale published his first paper on histamine **14**. This synthetic substance caused skin weals, irritation and,

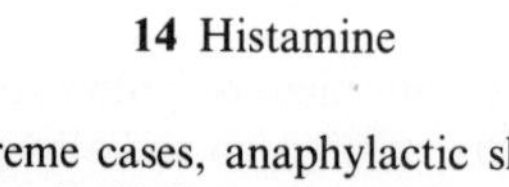

14 Histamine 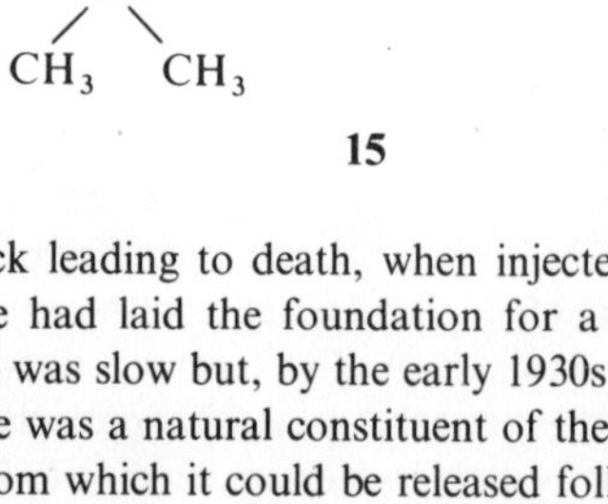   **15**

in extreme cases, anaphylactic shock leading to death, when injected into animals. Dale little dreamt that he had laid the foundation for a major advance in science. At first progress was slow but, by the early 1930s it was proved beyond doubt that histamine was a natural constituent of the body, that it was stored in specific cells from which it could be released following chemical or physical damage, and that many if not most hypersensitivity reactions were due to its release. The first to prove that it was possible to

The pharmaceutical industry today 15

antagonize histamine chemically was Bovet. He showed in 1937 that NN-diethyl-2-isopropyl-5-methylphenoxyethylamine **15** antagonized the histamine-induced spasm of various smooth muscles. The drug **15** was too weak and too toxic for clinical use, but its discovery demonstrated that an internally generated (endogenous), but sometimes harmful, chemical in the body could be antagonized by a synthetic material. Several dozen compounds now available are more potent and less toxic than the Bovet original but essentially doing the same job. One of these, chlorpromazine **9**, we have already discussed. This was introduced as an 'antihistamine' but is now, following chance clinical observation of its sedative properties, used as a major tranquillizer.

Drugs based on adrenaline. At about the time that Dale was determining the action of histamine he was also studying the action of a large number of synthetic amines related to the natural product, adrenaline **16**. This had somewhat earlier been isolated from the adrenals; its structure had been eluci-

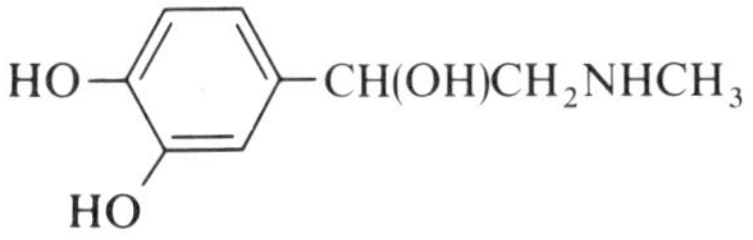

HO—⟨benzene⟩—CH(OH)CH$_2$NHCH$_3$ HO—⟨benzene⟩—CH(OH)CH$_2$NH$_2$
| |
HO   HO

16 Adrenaline 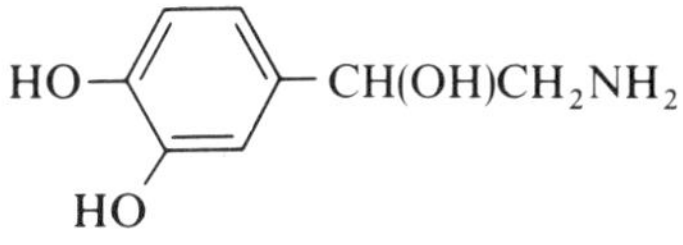   **17** Noradrenaline

dated and it had then been synthesized. Dale postulated that the sympathomimetic amines are responsible for the transmission of nerve impulses. Later it was found that both adrenaline and noradrenaline **17** were transmitters responsible for the impulses that maintain a stable blood pressure in man at rest and increase it in times of stress. In people with abnormally high blood pressure, it could be argued (and was), that an excess of these two transmitters was present in the system.

The biochemical synthesis of adrenaline (Scheme 1) starts from the natural amino-acid, phenylalanine, and proceeds by the aid of known enzymes through dihydroxyphenylalanine (dopa) and dihydroxyphenethylamine (dopamine) to adrenaline.

If then high blood pressure could be linked with an excess of adrenaline what more sensible than to block this synthetic pathway at some stage or other by a chemical. Such a chemical was **18** the simple α-methyl derivative of dopa.

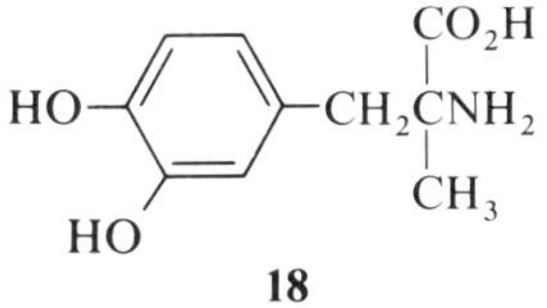

HO—⟨benzene⟩—CH$_2$C(CO$_2$H)(NH$_2$)CH$_3$
|
HO

18

This is taken up by the enzyme, dopa decarboxylase, just as if it were dopa itself and does, in part at least, decarboxylate to α-methyldopamine which in turn can be taken up by dopamine β-oxidase to provide α-methylnoradrenaline. At each stage, enzymes are blocked by the false substrate. Even when they unblock, they can do so only by producing another unnatural compound. In the event α-methyldopa has proved to be an effective, safe, and reliable antihypertensive.

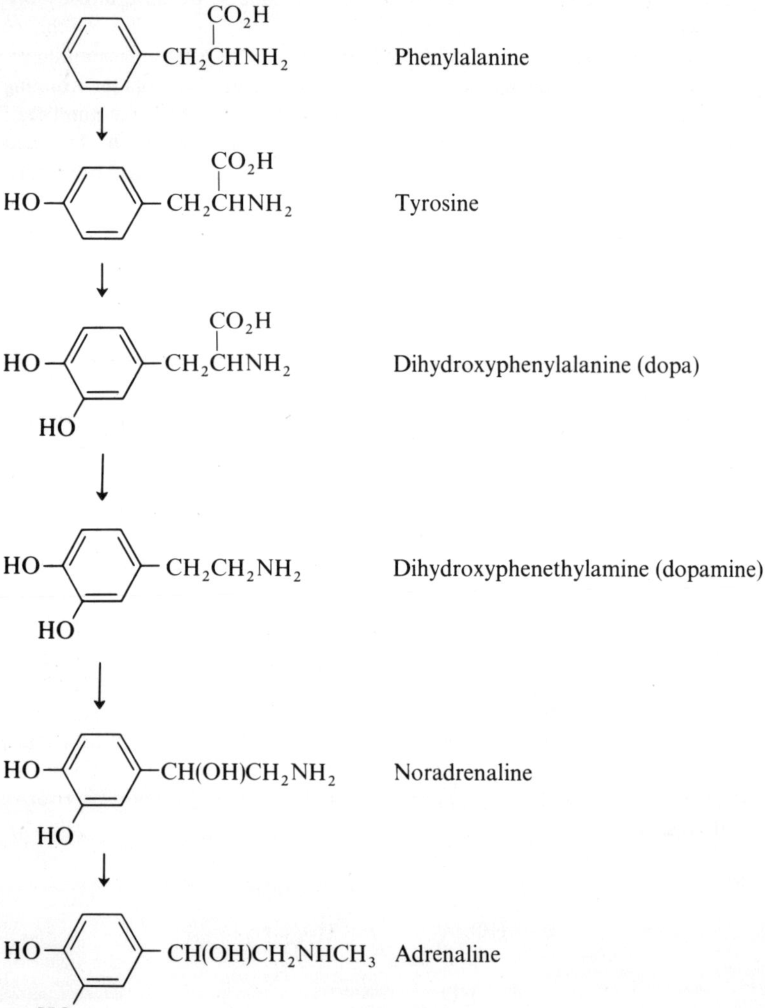

Phenylalanine

Tyrosine

Dihydroxyphenylalanine (dopa)

Dihydroxyphenethylamine (dopamine)

Noradrenaline

Adrenaline

Scheme 1

The pharmaceutical industry today 17

Folic acid antagonists. para-Aminobenzoic acid **19** was first synthesized in 1863, but remained of little biological interest until 1940. Woods and Fildes then showed that the sulphonamides act by competing with *p*-aminobenzoic acid, needed by bacteria in the synthesis of folic acid **20**. From this observation the concept of 'antimetabolites' was born. Since only bacterial cells make their folic acid in this way, while animal cells require preformed folic acid, the sulphonamides proved lethal to the bacteria but not to the host.

This observation was extended systematically by Seeger in 1947. He suggested that if a simple analogue of folic acid could be made it might be taken up in mistake by animal cells during rapid cell division, such as occurs in certain cases of cancer. Once the mistake was effected it would eventually destroy the multiplying cell. This hypothesis was soon tested. The first compound prepared was aminopterin **21** where a hydroxy function in the complex folic acid molecule had been replaced by amino. This produced dramatic though short-lived remissions in patients with leukaemia, and long-lasting remissions in the somewhat rare choriocarcinoma.

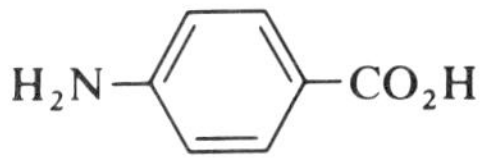

$$H_2N-\underset{}{\bigcirc}-CO_2H$$

19 PABA

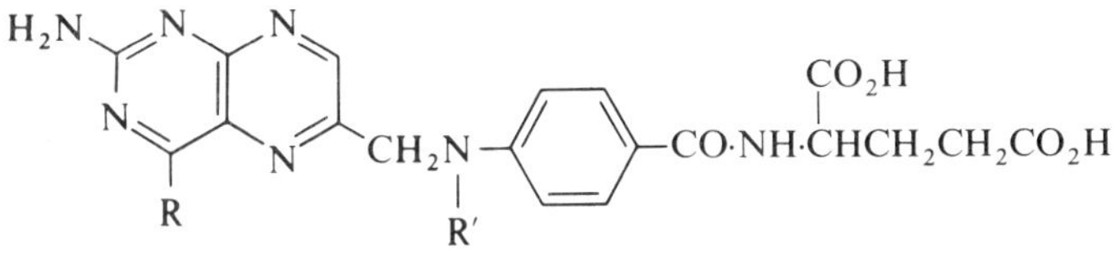

$$H_2N-\underset{R}{\overset{N}{\bigcirc}}-\underset{R'}{CH_2N}-\underset{}{\bigcirc}-CO \cdot NH \cdot CHCH_2CH_2CO_2H$$
$$|$$
$$CO_2H$$

20 Folic Acid, R = OH, R' = H
21 Aminopterin, R = NH$_2$, R' = H
22 Methotrexate, R = NH$_2$, R' = CH$_3$

The addition of a methyl group to aminopterin produced methotrexate **22** which was a superior drug in every way. This is now used routinely in cases of leukaemia (especially childhood leukaemia), in choriocarcinoma, and in severe psoriasis: remissions extending well over ten years have been achieved. More recently, the drug has been used in organ transplants since it inhibits the normal rejection processes of the body.

Field of science approach—physical chemistry

It is the dream of the medicinal chemist that one day, on the basis of existing knowledge, he should be able to tailor-make a drug to control a specific body function or to cure a disease.

Of course, the problem is exceedingly complex: drugs have to pass many barriers in the body before they reach their site of action. In the course of this they must not lose activity by metabolic attack yet they must eventually be eliminated from the body. At the same time as exerting their desired effect they should not produce undesired side-effects. These are all problems in dynamic equilibration. The physical chemist views the current 'hit or miss' approaches of the medicinal chemist as the blunderings of a man in a pitch-black room. Surely, he feels, it should be possible to systematize drug design? Work towards this end follows three approaches. One, the Molecular Orbital approach, seeks to determine the exact nature of the 'agonist' molecule exerting a deleterious effect in the body and by synthesis based on this understanding hopes to counteract its effect. This approach is far from certain—often the method hardly improves on the traditional 'hit or miss' approach. The second, or so-called Free–Wilson approach, relies upon a statistical analysis of the biological effects of several compounds; each change is given a numerical value and the sum of the effects is related to the sum of the various substituents present in each particular molecule. The third, and probably best accepted, approach is that of Hansch. Each of the methods depends heavily on computer analysis of masses of data but none more so than Hansch's approach. Measurements of various electronic (i.e. inductive and resonance) effects, partition coefficients between two solvents, ionization constants, and steric properties bestowed on the drug by various substituents are correlated with the dose of drug given and the response obtained. More than either of the first two methods, Hansch's approach measures those properties affecting the rate at which a drug actually gets to the site where it can exert activity.

So far no real success has attended the use of any of the three approaches; this application of science is still in its infancy. However, Hansch's techniques in particular are being increasingly applied to the problems of drug action and the first major advance achieved by their use is eagerly awaited.

3. Specific biological effects

J. F. CAVALLA

ONLY the briefest indication can be given here of the types of chemical compounds found useful in the cure, prophylaxis or amelioration of the many diseases of man.* Rather than give a mere catalogue of organic compounds of use in medicine, we have selected five examples to illustrate the fundamental thinking of the medicinal chemist and the factors influencing how he proceeds.

We have deliberately chosen two aspects of antibiotic therapy: penicillins and sulphonamides. The penicillins show how a chance observation led first to the discovery of a valuable drug and then, after many years of systematic modification, to more selective agents with essentially the same antibacterial action. On the other hand, the discovery of sulphonamide and its subsequent modification led not only to superior sulphonamide antibacterials but also to drugs of value in quite separate conditions. Our third example deals with hormones and serves to show how, by modification of natural compounds, more useful medicinals can be produced. The fourth section on antihypertensives shows how the first ideas for the treatment of high blood pressure stemmed from work designed for another purpose. Finally, the new science of tranquillizers or mood modifiers is discussed, wherein the team work of the medicinal chemist and the pharmacologist promises to bring under control some aspects of psychological abnormality.

The Penicillin antibiotics

The discovery of 'penicillin' by Fleming in 1929, followed by its isolation by Florey and Chain in the late 1930s, led to the marketing of two compounds in the mid 1940s. These were the so-called penicillin G **23** and penicillin V **24**.

$$RCH_2 \cdot CO \cdot NH - \underset{O}{\overset{}{\square}} \underset{N}{\overset{S}{\diagdown}} \underset{CO_2H}{\overset{CH_3}{\diagup}}$$

23 Penicillin G, $R = C_6H_5$
24 Penicillin V, $R = C_6H_5O$

Penicillins were made initially by surface fermentation in innumerable bottles and then, more efficiently, by deep vat fermentation. In one case **23**, phenylacetic acid was essential to the broth while in the other **24** it was phenoxyacetic acid. The two agents are roughly equivalent when injected but, by virtue of its better absorption from the gastrointestinal tract and superior stability in stomach contents, penicillin V is more effective than

---

* For an authorative and exhaustive account see Goodman and Gilman (editors), *The pharmacological basis of therapeutics*, Macmillan 1970.

penicillin G orally. Although other organic acids were tried in fermentation with the mould (*Penicillium chrysogenum*) used to make the penicillins, no superior product resulted until a team of chemists, biochemists, and microbiologists at Beecham Research Laboratories isolated free 6-aminopenicillanic acid (6-APA) **25** in 1958. Once this was available, making analogues of penicillin G and penicillin V became solely a problem for the medicinal chemist. 6-Aminopenicillanic acid is a complex dipeptide having free acidic

$$NH_2 \quad S \quad CH_3$$
$$\quad N \quad CH_3$$
$$O \quad CO_2H$$

25 6-APA

and amino-groups; only the amino-group has predictive utility. Treatment with phenylacetyl chloride and phenoxyacetyl chloride promptly gave penicillin G and penicillin V, respectively; derivatives of other acids gave analogues of penicillin. Since the discovery of 6-APA, thousands of semisynthetic penicillins have been made. The major objectives were two-fold: first, to produce penicillins capable of combating those organisms which produced penicillinase (which renders the drug inactive in the body) and, second, to produce derivatives which had a wider spectrum of activity against micro-organisms. In particular, compounds capable of eliminating gram negative organisms were sought and also those strains of bacteria which, though once susceptible, had become resistant to penicillin G. Other advantages sought were improved oral absorption, maintained and extended blood levels and reduced toxic hazard of the drugs.

The first semisynthetic penicillin to be marketed was phenethicillin **26**, a homologue of penicillin V obtained by treating 6-APA **25** with 2-phenoxypropionyl chloride. Although **26** is marginally superior to penicillin V by oral administration and gives higher blood levels, it did not accomplish the primary tasks outlined above. The first to do that to any degree was methicillin **27**, obtained from 2,6-dimethoxy-benzoyl chloride and 6-APA **25**. Methicillin is highly resistant to cleavage by penicillinase and thus, for the first time, a penicillin attacked bacteria which produce penicillinase. It is, however, poorly absorbed from the intestines and partly degraded by stomach acids; it therefore has to be given by injection. Cloxacillin **28**, on the other hand, was found resistant to cleavage by penicillinase and also orally active.

A major advance over penicillin G was ampicillin **29**. In this case, the acid used was D(−)phenylglycine. For the chemist it is of interest that the synthesis differs from that for other penicillins. The coupling acid, D(−)phenylglycine already contained an amino-group and thus had to be protected in one of the many ways open to the peptide chemist before coupling with

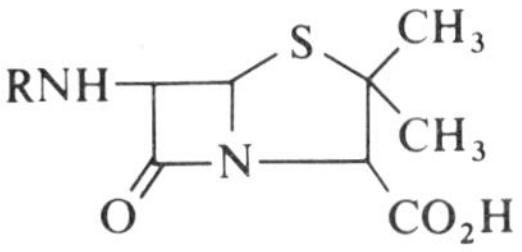

Specific biological effects

[Core structure: 6-APA penicillin nucleus with RNH-, S, C(CH₃)(CH₃), N, C=O, CO₂H]

26 Phenethicillin, R = C₆H₅–OCH(CH₃)CO–

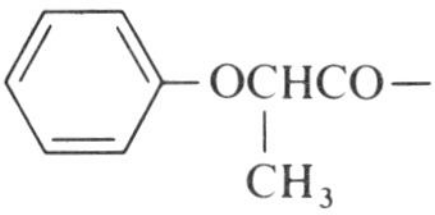

27 Methicillin, R = 2,6-(OCH₃)₂C₆H₃–CO–

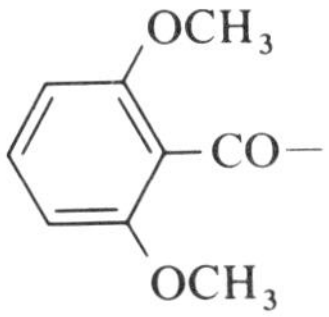

28 Cloxacillin, R = 3-phenyl-5-methylisoxazole-4-CO–

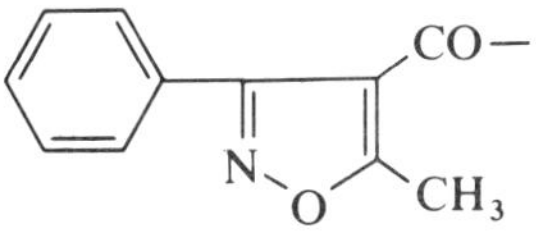

29 Ampicillin, R = C₆H₅–CH(NH₂)CO–

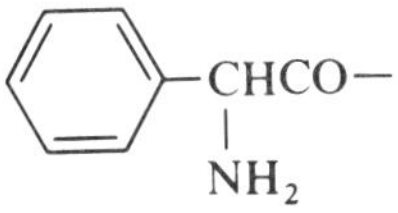

30 Carbenicillin, R = C₆H₅–CH(CO₂H)CO–

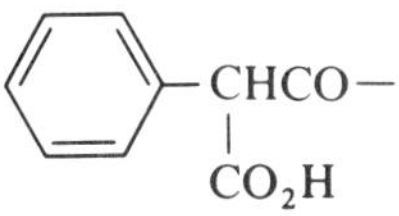

6-APA. Unlike methicillin and cloxacillin, ampicillin is not resistant to penicillinase. It is orally active. The reason for its dramatic success (world sales are now approximately £50 million annually) lies in its capacity to eliminate both gram positive and gram negative organisms. Only *Pseudomonas aeruginosa* and strains of *Proteus*, of the common bacteria, are not attacked; but with the passage of years, resistant strains of *E. coli*, *S. aureus* and *Aerobacter* are being increasingly encountered. For all that, in terms of benefit to mankind and general acceptance by a world market, few drugs have equalled ampicillin.

Finally, and in effect to round off the conquest of bacterial infection, carbenicillin **30** was introduced specifically for the treatment of *Pseudomonas aeruginosa* and those strains of *Proteus* resistant to ampicillin. Here, instead of an amino-acid, a dibasic acid was coupled with 6-APA in such a way as to leave one of the carboxy groups free. Unlike ampicillin, the drug has to be given by injection.

Thus, the vast majority of harmful bacteria can now be treated with one penicillin or another. But it is predictable that as the years pass, bacteria will acquire resistance to that particular antibiotic now used to eliminate them. The task of the medicinal chemist cannot therefore be said to be done. Constant attention must be paid to the dyke he has constructed; as the defences are breached by this or that organism, so further defences must be on hand. To this end, derivatives of 6-APA are still being made and, more recently, attention has been directed towards the synthesis of derivatives of 7-aminocephalosporanic acid **31** to give subtly different, though similarly acting penicillin analogues.

$$\text{31}$$

Sulphonamides

A discussion of sulphonamides as compounds having medicinal properties immediately reveals the value of looking at structures where activity is already known to reside. In the simplest sense, the fact that sulphonamide itself is active suggests that related compounds will, at the very least, be absorbed by the body, be transported to a potential receptor site and unless grossly changed, will have acceptable toxicity. On this reasoning many sulphonamides were made and tested experimentally. Some of the more valuable attributes of the series were discovered by physicians in giving the drugs to patients and examining their effects.

Sulphonamides as antibacterials

Before penicillin became generally available, the sulphonamides were the mainstay of antibacterial chemotherapy and contributed greatly to the sharp decline in illness and mortality caused by infectious diseases.

Initially, the chemical industry had been interested in the parent sulphanilamide **32** as a building block in the synthesis of azo dyes.

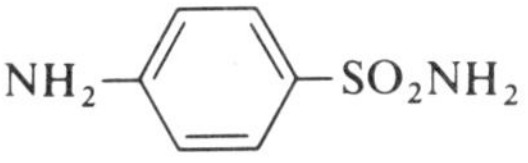

32 Sulphanilamide: $NH_2\text{-C}_6H_4\text{-}SO_2NH_2$

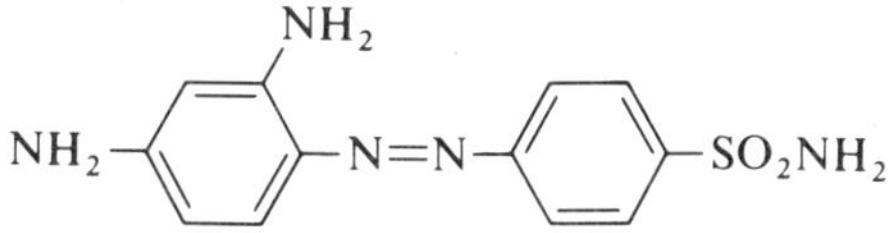

33 Prontosil: $NH_2\text{-C}_6H_3(NH_2)\text{-}N=N\text{-}C_6H_4\text{-}SO_2NH_2$

The antibacterial properties of these dyes had long been known but 'modern' medicinal research on sulphonamides began only when Domagk of I.G. Farbenindustrie showed in 1932 that mice with streptococcal and other infections could be protected by Prontosil **33**. The subsequent research work was, in retrospect, sporadic and ill-based. Much has been learnt about research methods in the interval. The real break-through for the medicinal chemist came in 1935 when Trefouels, Nitti, and Bovet observed that Prontosil was metabolized *in vivo* to sulphanilamide **32** and that this was as effective as Prontosil in curing experimental infections. In other words, the action of Prontosil in the body was probably due to sulphanilamide itself.

In the decade following this discovery more than 5000 sulphonamides were synthesized and tested but no more than a score were ever exploited commercially. One of the first of these was sulphapyridine **34** where the simple amide was elaborated to the pyridylamide.

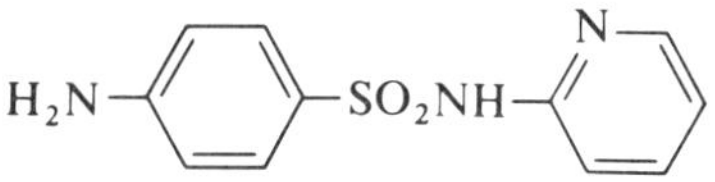

34 Sulphapyridine: $H_2N\text{-}C_6H_4\text{-}SO_2NH\text{-}(2\text{-pyridyl})$

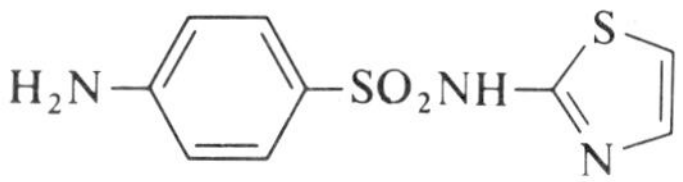

35 Sulphathiazole: $H_2N\text{-}C_6H_4\text{-}SO_2NH\text{-}(2\text{-thiazolyl})$

This, as M & B 693, made a major impact in the treatment of pneumonia. Sulphathiazole **35**, sulphadiazine **36**, sulphisoxazole **37** all followed, each an improvement on its predecessor. Further improvement was made by acylation of the free amino-group to give, amongst other compounds, phthalylsulphathiazole **38** which, by virtue of not being well absorbed from the

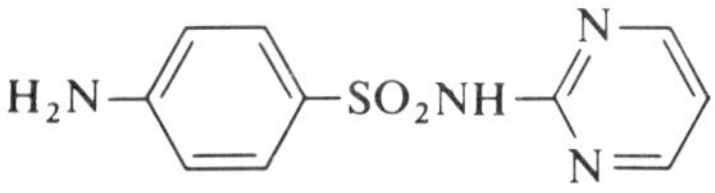

36 Sulphadiazine: $H_2N\text{-}C_6H_4\text{-}SO_2NH\text{-}(2\text{-pyrimidyl})$

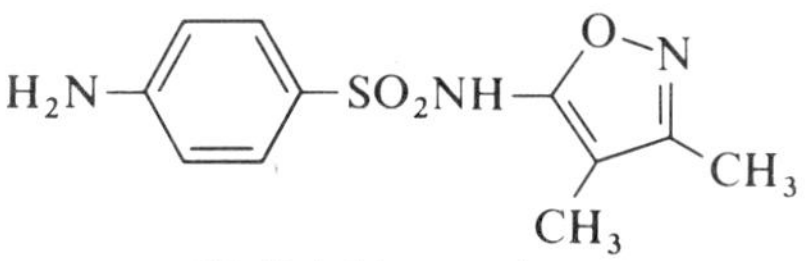

37 Sulphisoxazole: $H_2N\text{-}C_6H_4\text{-}SO_2NH\text{-}(3,4\text{-dimethylisoxazol-5-yl})$

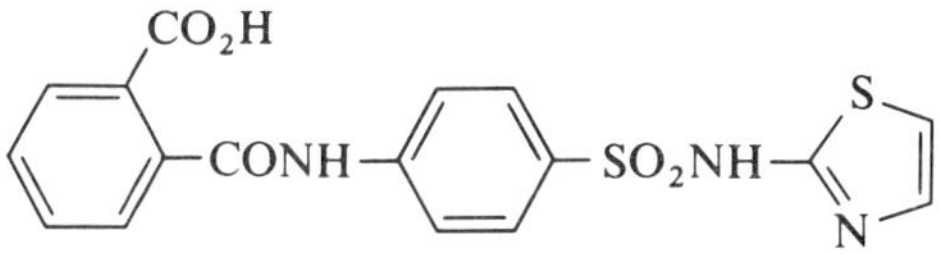

38 Phthalylsulphathiazole: $2\text{-}HO_2C\text{-}C_6H_4\text{-}CONH\text{-}C_6H_4\text{-}SO_2NH\text{-}(2\text{-thiazolyl})$

stomach, exerts its full effect in the bowel. Competition from the penicillins and other fermented antibiotics eventually limited the sales of the sulphonamides as anti-infective agents, but in recent years the combination of the sulphonamide, sulphamethoxazole **39** with the antimetabolite and antimalarial trimethoprim **40**, has provided a two-pronged attack against bacteria which is competing commercially on equal terms with the penicillins.

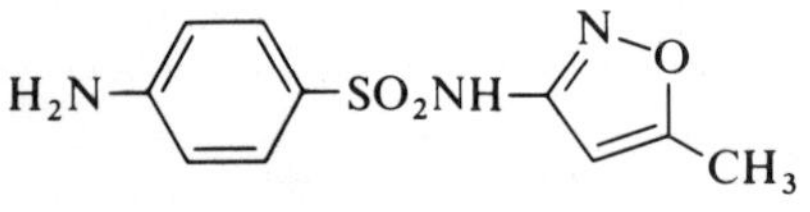

$H_2N-\langle\rangle-SO_2NH-\text{[isoxazole-CH}_3\text{]}$ $\text{[trimethoprim structure]}$

39 Sulphamethoxazole 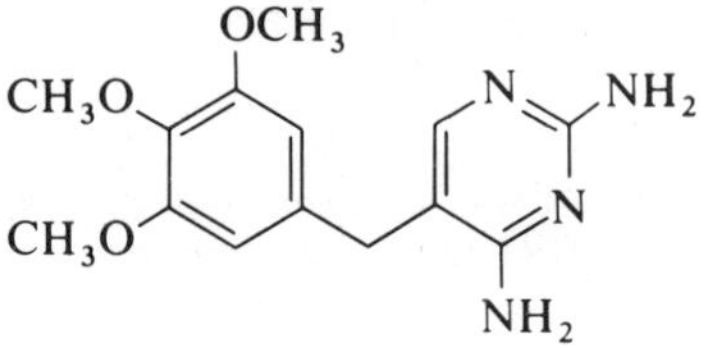   **40** Trimethoprim

Sulphonamides in the cure of leprosy

The discovery that sulphones can cure leprosy arose directly from the interest shown by medicinal chemists in sulphonamides. The compound, Dapsone **41** was made accidentally during sulphonamide synthesis. Fortunately it was tested at the time.

$H_2N-\langle\rangle-SO_2-\langle\rangle-NH_2$

41 Dapsone

It has generally weak antibacterial properties but some activity against tuberculosis. Somewhat later biological development work made it possible to screen compounds for activity against leprosy in rats when Dapsone was found to be the best available. It still is; all modifications of the structure have resulted only in, at best, marginal improvement.

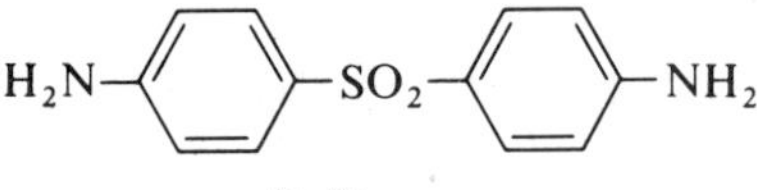

Sulphonamides as diuretics

Some years after sulphanilamide **32** was introduced as an antibacterial agent, the drug was found to inhibit the enzyme carbonic anhydrase. This enzyme is responsible for the degree of hydration of the animal system and has a direct bearing on the capacity of the kidneys to excrete unwanted water. The compound was therefore proposed as an oral diuretic at a time when the best method of treatment was with injected organic mercurials.

Since sulphanilamide itself was insufficiently active, medicinal chemists set about modifying it structurally. Several structure–activity correlations emerged quite different from those elicited in the antibacterial work. Substitution of the amidic nitrogen destroyed the *in vitro* carbonic anhydrase

Specific biological effects

activity; the benzene ring was far from optimal and a heterocyclic ring enhanced activity. As a result, acetazolamide **42** emerged as the first sulphonamide diuretic in 1950 and is still used today.

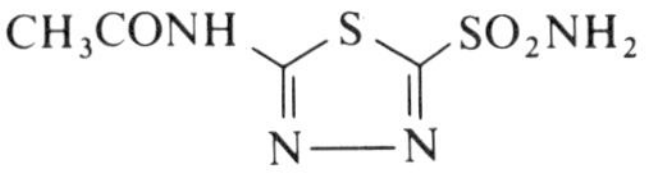

42 Acetazolamide

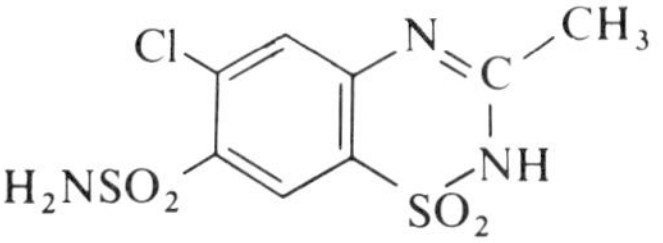

43 Chlorthiazide

While the relationship of acetazolamide to sulphanilamide is still obvious, that of chlorthiazide **43** is less clear. This compound arose from the observation that the presence of two sulphonamide residues improved the diuretic potency. It was then found almost accidentally that a compound such as **44** could react intramolecularly to give the benzothiadiazine ring system **43**. The resulting compounds not only possessed increased diuretic activity but also changed the character of the diuresis: the urine contained increased amounts of sodium and chloride ions with better retention of potassium by

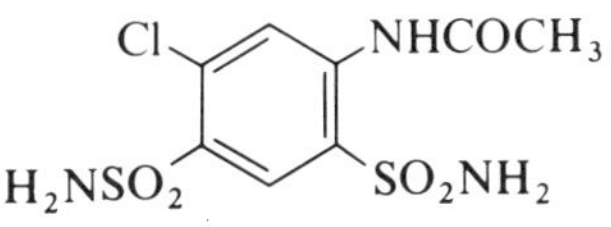

44

the host. Surprisingly, and unlike other sulphonamides, the benzothiadiazines do not inhibit carbonic anhydrase: their inherent activity is due to a direct effect on the renal tubular transport of sodium and chloride ion. Many analogues of the chlorthiazide structure **43** were made and permitted new structure–activity correlations. The double bond in the thiadiazine ring was unnecessary, the chlorine atom was better replaced by trifluoromethyl and large groups could be tolerated on the carbon atom in the heterocyclic ring. Incorporation of all these changes, provided the most potent derivative, bendroflumethiazide **45**.

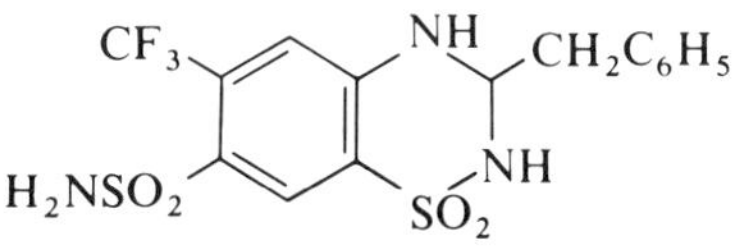

45 Bendroflumethiazide

At this stage, with an effective daily dose of only two milligrams, there was little point in continuing to search for more potent compounds. Yet it is not the potency alone that determines the usefulness of a drug. Within a few years a compound less potent than bendroflumethiazide on a weight basis was introduced. This, frusemide **46**, had the advantage of giving a greater

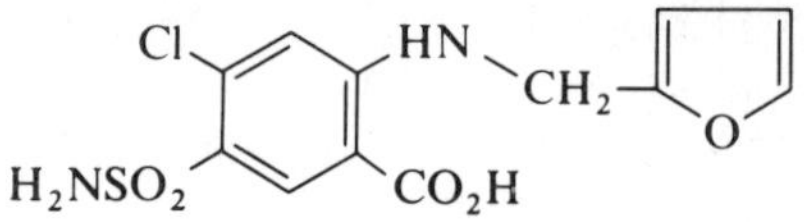

46 Frusemide

diuresis with a much prompter onset of action than any of the benzothiadiazines. Dramatic benefits may often be seen in oedematous patients within one half-hour of oral frusemide being taken.

Sulphonamides as hypoglycaemics
In 1942, certain French clinicians noted that patients being treated with an experimental sulphonamide for typhoid fever had reduced blood sugar levels. This observation was not followed up for over a dozen years until the antibacterial agent, carbutamide **47** was found to do much the same thing, possibly more dramatically.

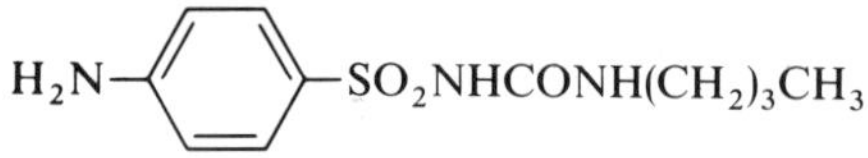

$H_2N-\langle\ \rangle-SO_2NHCONH(CH_2)_3CH_3$

47 Carbutamide

Experiment showed that the compound was acting by direct stimulation of the pancreas to produce more insulin and would therefore have predictable value in patients with diabetes. In the event, the compound proved useful only in the so-called maturity-onset diabetics; congenital diabetes failed to respond to oral medication and continued to require insulin injection. Nevertheless the amelioration of diabetes suffered by older patients as the pancreas ages is a target worthy of further research; once again the medicinal chemist modified the structure in the hope of getting better compounds. The outcome was the introduction of tolbutamide **48** and chlorpropamide **49**, both equipotent with carbutamide but less toxic. The chemists at Hoechst AG examined the basic phenylsulphonyl urea skeleton system-

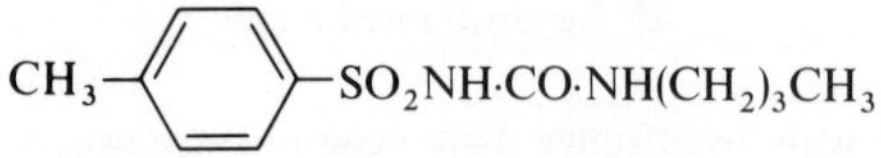

$CH_3-\langle\ \rangle-SO_2NH\cdot CO\cdot NH(CH_2)_3CH_3$

48 Tolbutamide

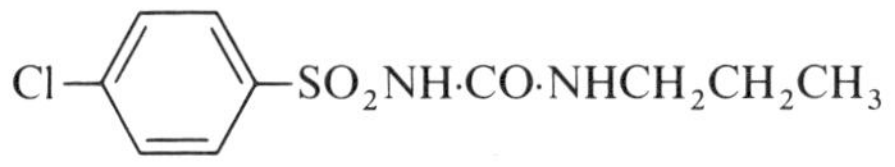

Specific biological effects

Cl—⟨C6H4⟩—SO₂NH·CO·NHCH₂CH₂CH₃

49 Chlorpropamide

atically and, following the synthesis and testing of over 500 compounds, finally showed that glibenclamide **50** was the most potent and the most free from side-effects of any substances in this series.

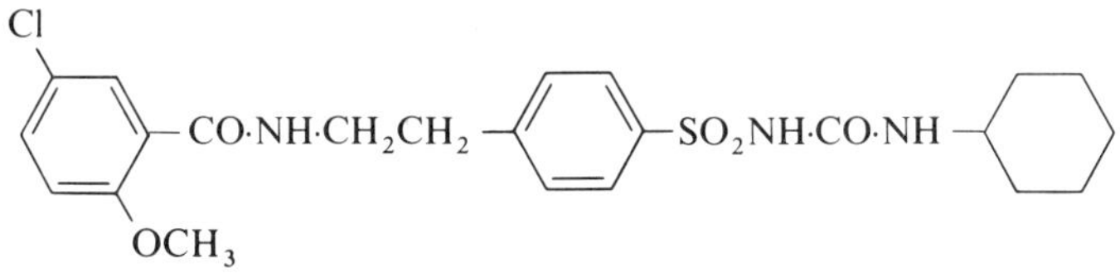

[Structure: 2-chloro-5-methoxybenzamide]—CO·NH·CH₂CH₂—⟨C6H4⟩—SO₂NH·CO·NH—⟨C6H11⟩

50 Glibenclamide

Hormones—Chemicals found in the body

The pharmaceutical industry does not rely wholly upon the manufacture of novel compounds capable of curing disease. It is much concerned with maintaining health. The human body is itself a miniature chemical factory daily synthesizing materials essential for a stable life. Any failure in this process can result in malfunction (mental or physical) leading to illness or death. It is possible in many cases to provide remedies by giving chemicals which promote or block the body's own chemical agents or stimulate failing body processes. The chemical administered may be the natural product itself, a compound imitating or convertible into the natural product, or an entirely different synthetic material.

Insulin—replacement therapy

A classic example of replacement therapy occurs in *Diabetes mellitus* where the pancreas fails to produce the agent necessary to metabolize the carbohydrate taken in diet. This failure resulted in death. The discovery by Banting and Best in 1921 that insulin was the agent involved was soon followed by the commercial isolation of pure insulin from the pancreas of the pig and sheep. Insulin is rapidly decomposed in the blood stream by proteolytic enzymes and treatment of patients with this material necessitated repeated injections. The contribution of the pharmaceutical industry was to prepare the more stable, longer-acting forms of insulin, specifically zinc insulin and protamine zinc insulin, both of which enabled the patient to maintain a steady metabolic state by only one injection per day. Although oral insulin-type agents (hypoglycaemics) are available for maturity-onset diabetes, for the deficiency disease made manifest in the young, insulin has to be used. In providing long acting insulin the pharmaceutical industry has accepted the role of modifying and purveying pure natural products.

The adrenals and their hormones

The physiological significance of the adrenals was first appreciated by Addison in the middle of the nineteenth century when he described the clinical syndrome resulting from destructive disease of the adrenal glands. The discovery of what exactly the adrenal or rather the adrenal cortex was excreting had to wait for another hundred years.

It was in 1948 that workers in America and Switzerland announced that cortisone and related structures are the major secretion of the adrenal cortex. At first, when crippled arthritics were treated with cortisone it was felt that a wonder drug had been discovered since dramatic remissions of their disease resulted. Unfortunately, the response was short-lived and in most cases the patients relapsed.

Of more long-term importance in human therapy was the opportunity presented by the discovery of cortisone for the treatment of adrenocortical insufficiency (Addison's disease). This is another obvious case of replacement therapy. However, cortisone did much else besides, despite its failure in arthritis. This failure has resulted in massive research programmes to modify the natural hormone so that its benefits may be gained without the slow cumulative toxic effects of the natural hormones. Even for the 'rational' patients with impaired or non-existent adrenals, semi-synthetic forms of cortisone are now preferred to the natural hormone.

The corticosteroids have many different physiological actions. They influence the metabolism of the food we eat, they regulate the excretion of electrolytes and water from the body, they act on the cardiovascular system and on the central nervous system, and, in times of stress and infection, they mobilize body resources to counter the affliction. Without them, life would be possible—but only just. This multiplicity of actions affects the dose at which the compounds are prescribed: for those without adrenals, but healthy otherwise, a small dose suffices. For the sick, quite large doses may be neces-

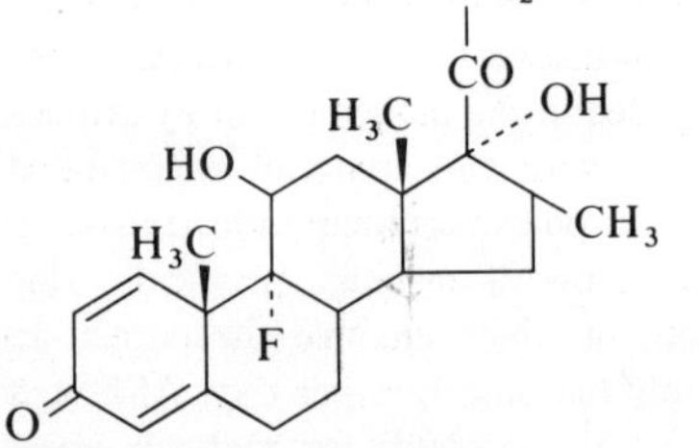

51 Betamethasone 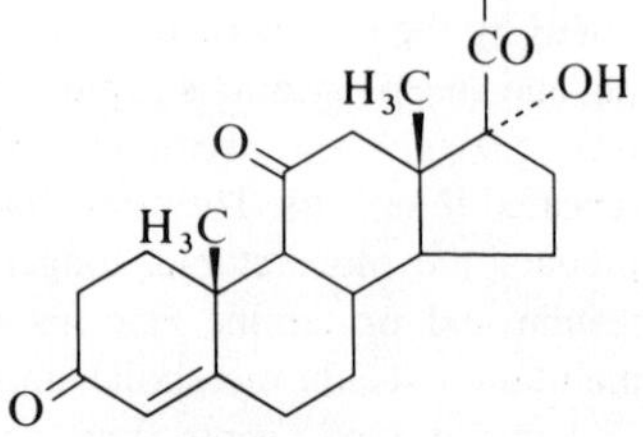   **52** Cortisone sary. The semi-synthetic 'improved' cortisones have the advantage of selectivity of action so doses can be given for specific effects without harm to the patient. A comparison of the very potent betamethasone **51** with the natural cortisone **52** provides an example. The skeleton of the compounds is basically the same, only the substitution pattern differs. In cortisone, beneficial liver glycogen deposition and anti-inflammatory effects go together with the undesirable sodium retention properties; with betamethasone these actions have been separated.

Other steroid hormones—fertility control

The sex hormones are other naturally occurring substances widely studied by the pharmaceutical industry. Progesterone **53** and oestradiol **54** are both intimately concerned in the maintenance of a fertile state in females. An excess of either at certain times of the oestrus cycle can prevent ovulation or conception. This may be desirable or undesirable for the female. In the early days of the research on oral contraception, it was felt that these compounds themselves were potentially useful for this purpose, but it was soon shown that the doses necessary for ovulation suppression or fertilization control

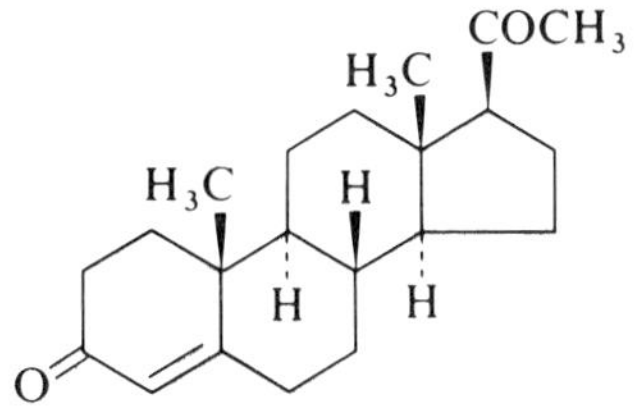

53 Progesterone

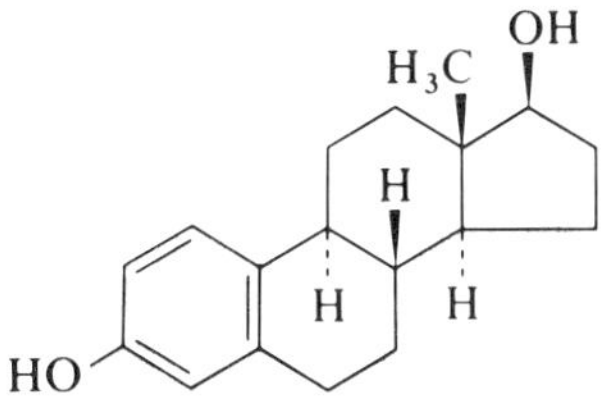

54 Oestradiol

were too great. At these doses, the undesirable side effects proved too severe.

Attempts were made to modify the structures and separate the desirable effects from the undesirable. This work soon proved fruitful and many novel compounds were introduced by various companies. On a strict weight–potency basis, the most active progestin is norgestrel **55** which is active at 0·5 mg per day. This compound is of particular interest since it is completely

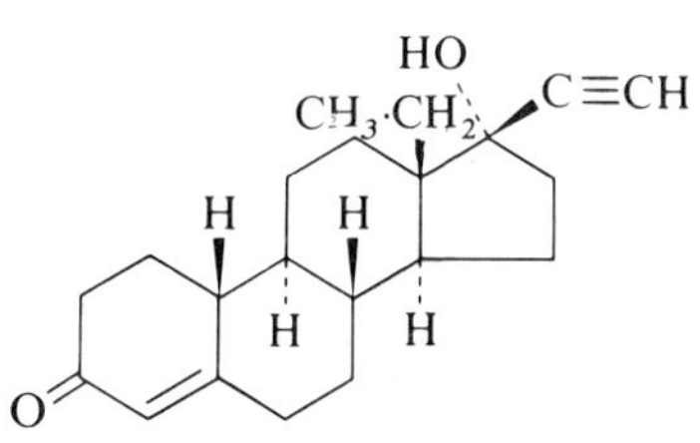

55 Norgestrel

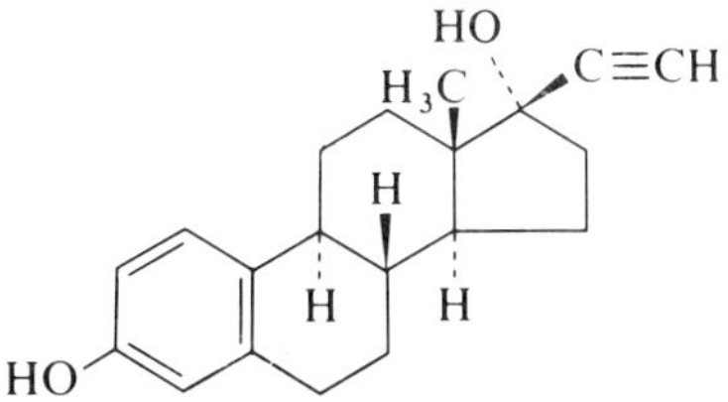

56 Ethinyloestradiol

synthetic, unlike all the other commercially available progestins whose basic skeleton is derived from a natural compound. The oestrogen which proved most active was ethinyloestradiol **56** which is capable of exerting its effect at only 0·05 mg. Combinations of agents of this type now offer the greatest hope of a rational solution of the problem of population control.

Work is continuing to find other hormonal means of preventing conception. Transport of the male sperm through the vaginal cervix can now be prevented by agents based on semi-synthetic progestins. With these, the normal ovulatory cycle of the woman can be maintained. Another technique is to modify the hormonal balance so that fertilization of the ovum is permitted, while the implantation of the resulting blastocyst in the wall of the uterus is prevented. This is the basis of the postcoital or 'morning after' pill.

On the horizon we can discern the strong possibility that in the 1980s a prostaglandin, or more probably a semi-synthetic prostaglandin analogue, will be commonly used in fertility control. The prostaglandins were discovered in the 1930s during work on the constituents of human seminal plasma. The complex structure, not unravelled until twenty years later, is based on a complex fatty acid. The structure is markedly specific and most attempts to modify it result in complete loss of activity.

Prostaglandin $E_2\alpha$

Exactly what the prostaglandins do is still under investigation: certain of their actions are well described. Of clear interest is their ability when injected to induce the birth of a baby normally at term. Of possibly greater value, is their ability to induce an abortion when given intravaginally at any period up to three months following conception. From this it will be seen that we have the making of a 'once a month' suppository or tampon which can induce menstruation irrespective of whether the woman is pregnant or not. Such a product may well be thought undesirable by certain ethnic or religious groups today, but in the coming years it seems certain to require fuller consideration.

The prostaglandins in their natural form are inactive when taken by mouth and structural modification will therefore be necessary. Once an oral form is produced, then another method of contraception based on compounds naturally occurring in the body will be available.

Minor tranquillizers

Mention has been made earlier (p. 10) of the discovery of the tranquillizer, chlordiazepoxide **6**. This was made by treatment of the quinazoline 3-oxide **57** with methylamine. Had aniline been used instead of methylamine the series might never have been discovered: with aniline the rearrangement

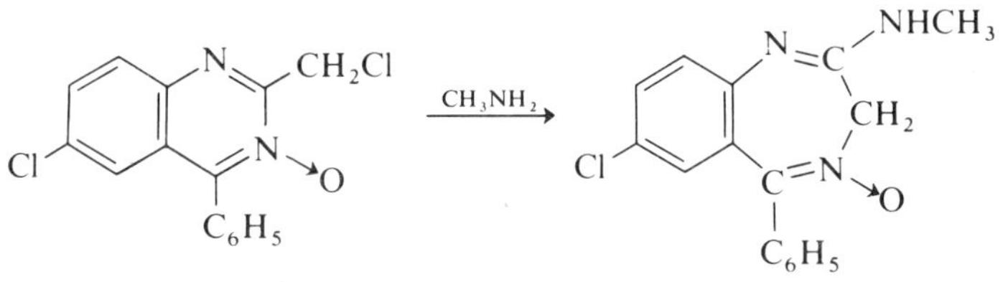
CH₂Cl
CH₃NH₂
NHCH₃
CH₂
Cl
Cl
C₆H₅
C₆H₅

57 **6** Chlordiazepoxide

does not occur and only the predictable substituted quinazoline **58**, devoid of sedative properties, is formed.

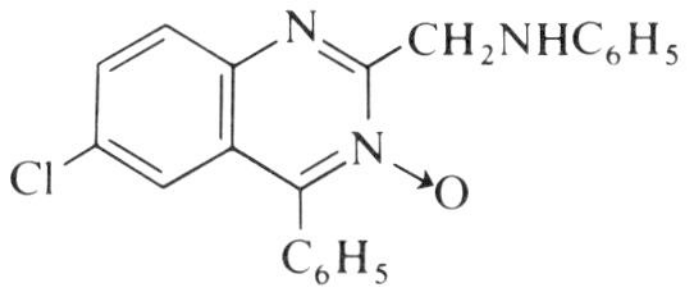
CH₂NHC₆H₅
Cl
C₆H₅

58

At the time of the discovery of the benzodiazepines, meprobamate **59** was the tranquillizer of choice. Modification of the meprobamate structure

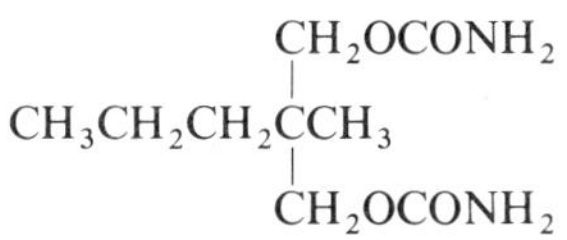
CH₂OCONH₂
CH₃CH₂CH₂CCH₃
CH₂OCONH₂

59 Meprobamate

failed to yield improved compounds and few related structures are now used clinically. Chlordiazepoxide, however, possessed so novel a structure that this fact alone stimulated interest in modifying it. In the event, many related compounds were found to have sedative properties of differing degrees and quality.

Chlordiazepoxide **6** is an *N*-oxide and contains a ring amidino-group. Reduction of the N$\rightarrow$O grouping and replacement hydrolysis of the amidine gives a more active compound, diazepam, **60** which is now

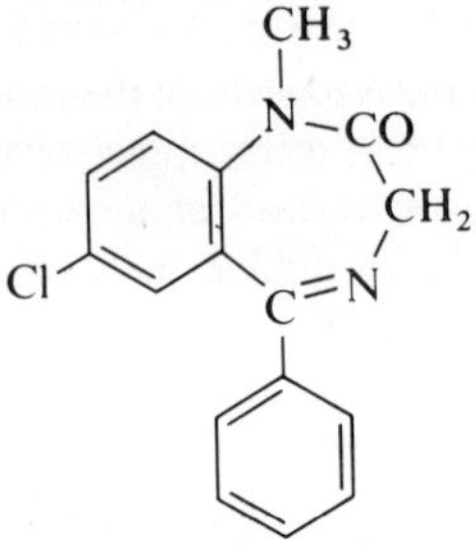

60 Diazepam

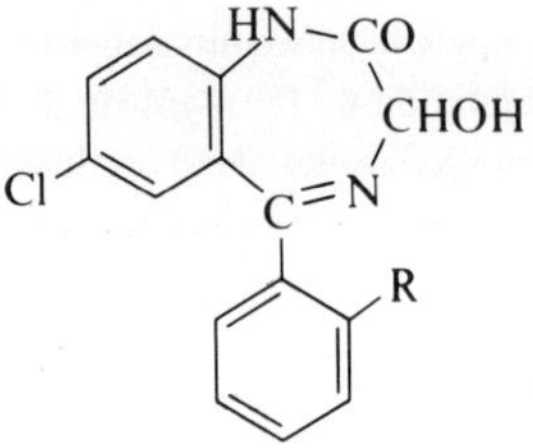

61 Oxazepam, R = H
62 Lorazepam, R = Cl

steadily replacing chlordiazepoxide as the world's most popular prescription drug. (Sales in the U.K. alone in 1971 were over £5 million). Diazepam is metabolized by man in part to oxazepam **61** and this too is now available commercially. The introduction of another chlorine atom, this time into the unsubstituted ring, gives lorazepam **62** more potent even than diazepam and with a similar activity.

The potency of these compounds is measured in the laboratory by observing the degree of muscle relaxation they cause in mice, by their capacity to tame monekys, and to inhibit fighting in certain rats with brain damage in the septal region of the brain. The combination of these results gives a measure of the anti-anxiety effect of the compound, which has then to be compared with its overall sedative activity. It is of little use alleviating anxiety in man if the result is a completely comatose population. The advantage of the benzodiazepines lies in their ability to alleviate anxiety without causing an unacceptable degree of sedation.

Further modification of the diazepine structure produced nitrazepam **63** where a nitro-group replaces the chlorine atom in the benzene ring. Unlike

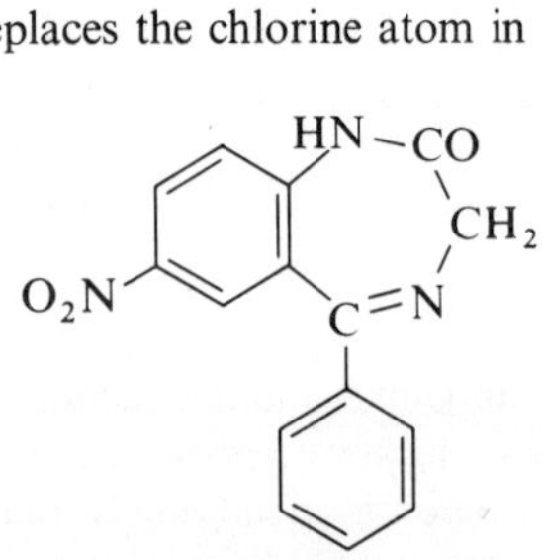

63 Nitrazepam

diazepam, nitrazepam has less anti-anxiety action but greater sedative–hypnotic properties: it is therefore used to induce sleep. There is no obvious reason why such a modification should produce so different a profile of activity. This finding thus demonstrates most clearly that one task of the research team is to synthesize and test 'likely' structures that will provide wider knowledge. The team must use whatever philosophy or hypothesis they can, not merely test mechanically for one activity in the hope of finding the 'ideal' compound in that series.

Major tranquillizers and antidepressants

Chlorpromazine **9** was introduced by the French company, Rhône Poulenc, as a compound capable of potentiating the effects of anaesthetics and inducing artificial hibernation. It followed from the discovery that earlier phenothiazine derivatives, themselves introduced as antihistamines, had this property. Once used in man, the drug was shown also to induce a sense of withdrawal in agitated patients and this led to its use in major psychotic states.

Its success was immediate: it is estimated that between 1955 and 1965 more than 50 million patients received treatment with chlorpromazine and more than 10 000 publications dealt with its actions. The accumulated evidence shows that it causes psychotic signs in the majority of patients to lessen to an extent which in many can be classed as a cure. Moreover, the diminution of bed occupancy of psychiatric hospitals which has occurred in all Western countries over the last fifteen years is due in major part to the use of the drug. The resulting economic gain is immense.

The motivation for the medicinal chemist to build and improve on a success of this order is clear. Variations of the phenothiazine ring, of its substituents, and of the basic side-chain have been made in profusion: certainly many thousand chlorpromazine-like structures have been tested. Yet only marginal improvements have resulted from all this effort. Chlorpromazine remains the compound of choice in all major psychotic disorders and is likely to remain so for many years.

However, in the course of this work riches of another kind were found. Structures related to chlorpromazine, the first of which was imipramine **64**, were shown to have remarkable properties in alleviating the malaise of

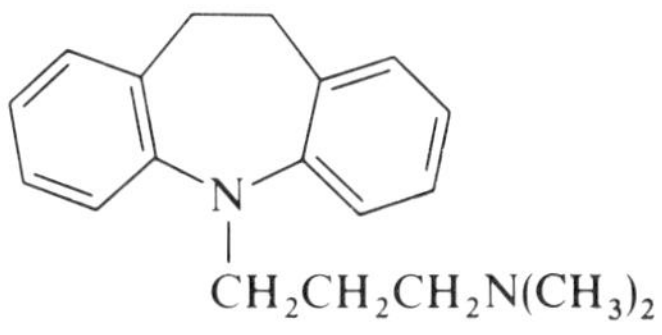

$CH_2CH_2CH_2N(CH_3)_2$

64 Imipramine

certain depressed patients. In the first controlled clinical trial of the drug, clear benefit was derived by patients suffering from so-called 'endogenous' depression characterized by inactivity and withdrawal. Hyperactive, agitated, and anxious depressives were made worse by the drug.

The search for structures chemically related to imipramine has been intensive. Several have been marketed: some are more potent and others more selective yet none is convincingly superior to imipramine itself. In the field of antidepressants, as opposed to the major phenothiazine tranquillizers, our best hope is that the chemist in alliance with the biochemist working on the uptake of brain amines, may yet find an 'optimum compound'.

Antihypertensives

The reason for, or causes of, most cases of hypertension are not understood. Yet hypertension is one sign of a circulatory disorder forming a major cause of death. Much work has therefore been done to produce compounds capable of lowering blood pressure, even though such work has, of necessity, been empirical. For guidance chemists have used objective theories even when their scientific rationale is sketchy.

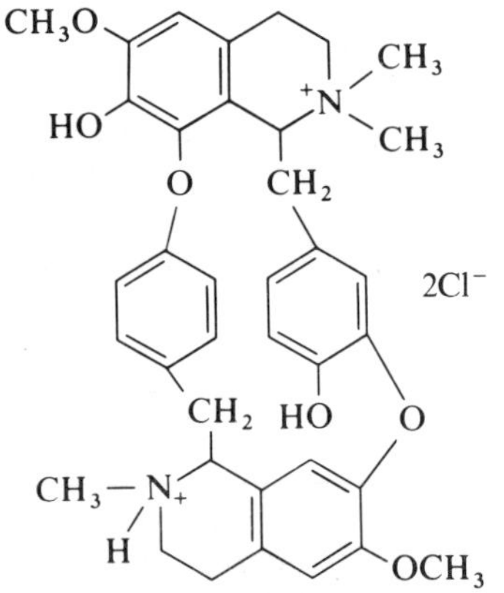

65 d-Tubocurarine

$(CH_3)_3 \overset{+}{N}(CH_2)_n \overset{+}{N}(CH_3)_3 \quad 2Br^-$

66 Decamethonium, $n = 10$
67 Hexamethonium, $n = 6$

The muscle paralysant d-tubocurarine **65** was believed to contain two quaternary ammonium groups separated in space by 1·2 nm. To determine the effect of separation of the nitrogen atoms on activity, Barlow and Ing made a series of simple quaternized $\alpha\omega$-diamines with varying numbers of $>CH_2$ groups. One of these, the so-called decamethonium **66**, had neuromuscular blocking properties similar to the natural curare derivative. In this work the derivative containing two quaternary groups separated by six methylene groups, the so-called hexamethonium **67**, was found to possess potent ganglion blocking action. Blocking ganglionic transmission of nerve impulses causes a drop in blood pressure. Thus hexamethonium found a use in hypertension.

Specific biological effects

While the ability of hexamethonium to reduce blood pressure was indisputable, it had several undesirable features, including poor oral activity. This was more than enough to encourage the search for better compounds. In the mid-1950s mecamylamine **68** and pempidine **69** were found, both were orally active and more certain in their action than hexamethonium. Even so, they still had many side effects, including constipation, impotence,

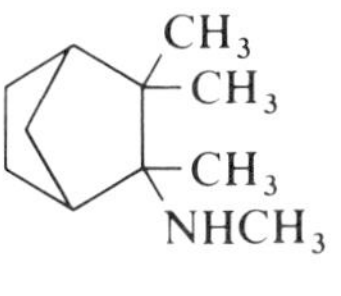

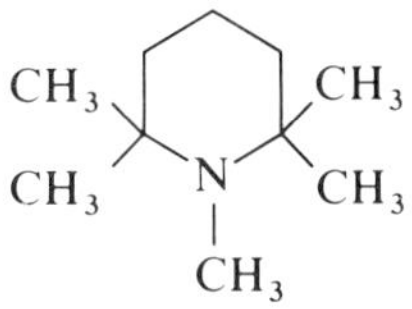

dry mouth and blurring of vision. Pharmacologists suggested that a more selective attack than blockage of all ganglionic impulses, namely blockage of the adrenergic neurones, could lower blood pressure with fewer side-effects. Once this possibility was recognized a number of research groups found similar but different drugs expressly for this purpose: bretylium **70** guanethidine **71** and bethanidine **72** of which the last two are still much used.

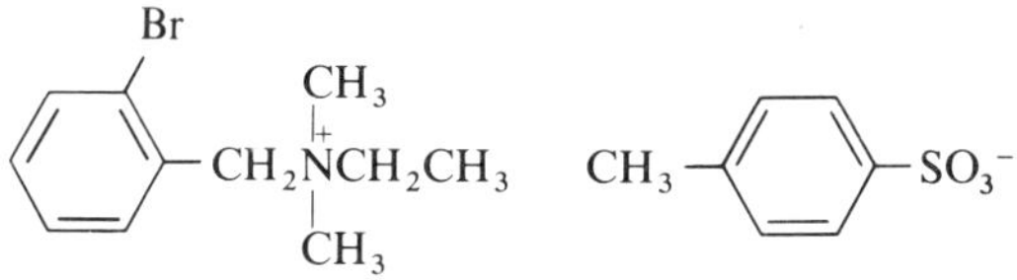

70 Bretylium

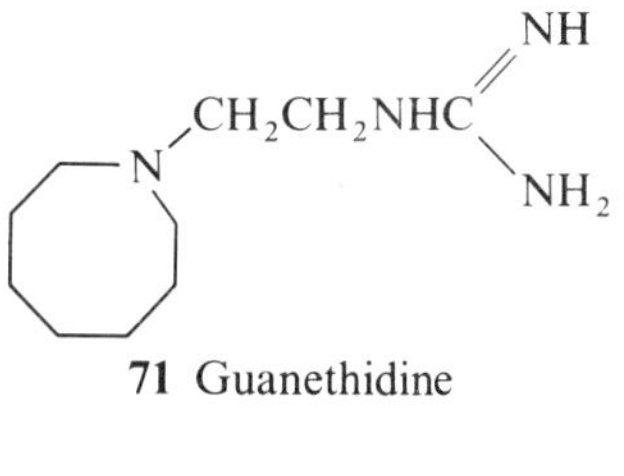

71 Guanethidine

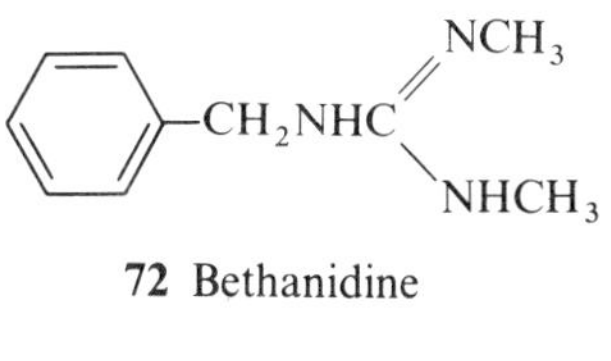

72 Bethanidine

A completely different approach to the treatment of hypertension is described in an earlier discussion on α-methyldopa (p.15).

4. The industrial development of a new drug J. F. CAVALLA

Team-work

THE medicinal chemist has had many successes in the conquest of disease, but it would be inaccurate and unfair to suggest that he is the 'onlie begetter of these sonnets'. Long before the chemist arrived on the scene, natural products were being used on a trial and error basis. Today, the chemist working alone would have no chance whatsoever of finding drugs better than those we have already. He needs partners. His first must be a biologist —in most cases a pharmacologist. Any decision to separate chemist from pharmacologist, even if only by placing them in separate buildings, is ill-advised. The two work as a team, the absence of either causing ruin to the design. While the routine pharmacological task is to assess the activity of the compounds supplied by the chemist, a slavish adherence to this procedure will result in the loss of many valuable discoveries. The competent pharmacologist should be able to advise the chemist that, while his compound has little potential in market A, i.e. it shows little activity against disease A, it may find a place in market B since it shows some activity against entity B, although the activity will have to be modified in certain ways. Such advice need not rest entirely on objective results if the two scientists are working as a team. An intuitive 'feel' or 'hunch' for the subject can be explored in this way without recourse to formality; purely subjective enthusiasm may then open new exploratory experimentation.

Screening of chemicals

There is also work which the pharmacologist will do alone or allied with scientists (biochemists, clinicians etc.) other than the medicinal chemist. This must be so in the design of models or 'screens' for assessing drugs in complex biological systems or disease states. These models are increasingly important to progress in medicinal practice. Thus it is not wholly because of the difficulty in finding suitable chemicals for the purpose, that a cure for cancer has not been found. A cure for *mouse* cancer has already been found: the difficulty lies in finding a test-bed, i.e. in persuading small animals to contract tumours which are non-specific to them but directly relevant to the human condition. Such reliable models, when obtained, will bring a solution of the problem of cancer within sight. Testing problems also dominate the search for a cure of viral disease: the discovery of compounds specific for one virus alone is of little help because of the multiplicity and complexity of viral species and strains.

Degenerative heart disease illustrates another classical situation urgently requiring valid drug therapy where absence of a reliable animal model useful

The industrial development of a new drug 37

for screening large numbers of compounds is causing delay. In certain heart conditions, artificial animal preparations do already exist and are of some value but the overall progress is slow.

In mental illness, the discovery of chlorpromazine by accident in the clinic should have resulted in the extrapolation back to an animal model. But, so far, all successes have depended on essentially artificial test-beds. No true 'schizophrenic mouse' has been obtained against which future drugs can be evaluated.

Safety of drugs

Before the best compound in a series is found, the toxicologist is called in to assess the safety factor of the drug or the series. In most cases the best compound will be that with the highest therapeutic index, i.e. the ratio of lethal dose to active dose. The most potent compound is not always the best; frequently a less active but also less toxic congener will be chosen. Once chosen, the compound is subjected to the most severe safety evaluation that can be devised. As our knowledge advances, stringency of such tests increases. It is now conventional to use two animal species one of which is a rodent, and to dose them exhaustively seven days a week for at least six months at three dose levels. One of these levels is a deliberate overdose so that clear damage is inflicted on the animal. The experience so gained is then used to show that at the lower doses no damage, least of all that demonstrated at the top dose, is encountered. The animals are exhaustively examined both macroscopically and microscopically and weight gain and blood picture are compared with controls.

When the drug is intended to be given for long periods, as for instance in antihypertensives or hypoglycaemics, six months of toxicology are inadequate and the term is extended appropriately. For oral contraceptives, for instance, toxicity trials are under way in dogs and monkeys and scheduled to last for seven and ten years, respectively. The pharmaceutical industry does not dispute the demand that it should do every conceivable test it can to ensure safety; but at times new dangers become apparent at a later date which were not foreseeable when the drug was first tested.

This happened tragically with the drug thalidomide **73** which was introduced as the first wholly safe hypnotic. This claim was made because it had

73 Thalidomide

proved impossible to kill rats with the compound, at a time when barbiturates were causing regular deaths from overdosage. Thalidomide was widely used as a sleeping pill. It also proved of value in anxiety states and sleeplessness in pregnancy. The defect of thalidomide lay in its unsuspected power to affect the development of the unborn foetus and thus to permit or cause the birth of deformed children. At that time, this was a side-effect not previously encountered in drug therapy and not even believed possible. Once it had been shown—by statistical correlation techniques—to occur in humans, the effect was looked for and found in animals. As a result a whole new science, called *teratology*, was born, and all candidate drugs for pharmaceutical use are now submitted to such tests.

When thalidomide is given to pregnant rats it is possible to demonstrate the production of foetal abnormalities, but only with difficulty; with laboratory rats as the sole species, this work would show aspirin to be as teratogenic as thalidomide. However, since thalidomide is clearly more teratogenic than is aspirin in humans, other, more reliable, species were searched for and found. The best for this purpose proved to be the rabbit which showed the effects of thalidomide exceptionally well, while aspirin produces few rabbit abnormalities. The monkey is also susceptible to thalidomide but with its extended period of gestation does not provide a sensible species for investigation.

As a result of this work, new requirements were laid down: for a new medicine to be adjudged free of teratogenic risk it should produce no abnormalities at maximum dose tolerated without ill-effect in both the rat and rabbit. Certain countries, especially those European ones which suffered most from the effects of thalidomide, make assurance extra sure by incorporating the mouse as a third species.

While the production of a deformed child can be shown, with difficulty, to have been caused by the taking of a certain compound many months earlier, the possibility of one's grown children themselves actually producing abnormal offspring as the result of their parents' drug-taking habits is even farther removed from detection. Yet the possibility is there and is being guarded against as well as possible by fertility studies and by examination of the chromosome pattern (specifically of the Chinese hamster) of litters from drug-treated parents.

In every pharmacological and toxicological experiment, the assumption is made that what happens in the animal will predictably occur in man. This may not necessarily be so, and to guard against preventable toxic hazard the metabolism of the compound by test animals is compared with that shown by human volunteers. This is a complex and difficult procedure requiring much expensive equipment, such as liquid scintillation counters for measuring isotopically labelled materials and mass spectrometers to help in determining the structure of products of metabolism. The isolation of

The industrial development of a new drug 39

metabolites from urine, faeces, blood, and tissue is possible now only by the use of physical techniques unknown a dozen years ago.

Once identified, the metabolites are themselves tested as active drug substances and, on occasion, have supplanted the drug as the agent of choice: oxyphenbutazone **74**, identified as the metabolite of phenylbutazone **7**, is widely preferred to its precursor.

74 Oxyphenbutazone, R = OH
7 Phenylbutazone,   R = H

Clinical trial

Once the toxicologists have agreed on the safety of a compound, the next step is its clinical trial. The standard method now used is the 'double blind study' in which the test drug is compared with either a placebo or standard in forms indistinguishable from one another. Not only is the patient unaware of which compound he is taking but so also is the physician whose responsibility it is to evaluate the effect. By means of this technique, any subjective impression the patient or doctor may have regarding the merits or demerits of a particular drug is removed and an objective assessment is obtained. This is important. The 'placebo' effect of an inert substance can be quite profound: lactose for instance is truly inert. Yet it can appear not only to act as a pain reliever but also to affect the autonomic nervous system. When a confident physician prescribes this 'drug' to a patient who wants to get better, the blood pressure actually drops.

Formulation

During the investigation by the clinical pharmacologist, the drug-house pharmacist has been perfecting the form the drug will take when it eventually comes to be marketed. There are many possible forms: from the simple white tablet to the sophisticated micro-encapsulated slow-release multicoloured formulation designed with both appearance and effectiveness in mind. Three things will be uppermost in the pharmacist's mind: purity, stability, and availability of the drug at the site of action.

Purity requires analytical precision coupled with quality control. The drug substance will, of course, have to comply with an analytical specification set by the pharmacist but in many tablets the drug substance is only a fraction of the whole. Moreover, with the ever increasing potency of drug substance, this fraction can be microscopically small. 'Excipients' make up the rest. In a recently introduced treatment for migraine, the quantity of active material in each tablet is only 25 micrograms or, to put it another way, 25 grams is enough for a million tablets. With such quantities, the control measures needed to ensure exact dosing are both complex and costly.

Stability is measured both naturally under ambient conditions and in an accelerated manner using high temperatures and humidity. To repay the considerable expenditure on R & D, any proven drug must be acceptable world-wide. If it is exported, its stability must be predictable even in the most adverse conditions in the worst climatic zone.

Finally, bio-availability is of increasing concern to the pharmacist as drug potency and effectiveness increase. At one time it was believed that if you took a tablet you got the medicine. This now has to be proved, usually by blood-level estimations in volunteers, or, less satisfactorily, by dissolution measurements in simulated stomach fluids. Clearly, the coating of a tablet can influence drug availability. In one case, where an enteric coated form of aspirin was marketed with the aim of preventing absorption of the drug in the stomach (where it might well cause ulcers in sensitive patients) but allowing it in the intestines, it was found that many of the tablets were passing through the body completely unchanged. In another, more subtle, example the particle size of the actual drug (the antifungal agent, griseofulvin) was found to have a profound effect on its solubility, with the result that the dosage could be halved when the drug was micronized. It is this aspect of pharmaceutical technology which advises caution when attempts are made to utilize cheaper generic forms of proprietary drugs. One may pay less for the medicine but achieve nil effect as a result.

Recent advances in pharmaceutical engineering have provided means of administering drugs routinely which were once impossible. Metered aerosol canisters will deliver a measured dose of drug direct to the lung; elegantly designed insufflators will allow solid drug to pass through nose or mouth, again to the lung. Long term treatment can be effected by injection of 'depot' forms of the drug. Drugs, such as oxytocin, which were believed active only when injected can now exert their effect in special tablets designed for buccal absorption. So-called 'positioned-release capsules' can be used to allow release of the drug only in the alkaline condition of the duodenum while the preparation of relatively insoluble forms of the drug substance will provide for maximum effectiveness in the bowel. We have recently seen the emergence of a drug company whose sole purpose is to provide superior methods of drug application as close as possible to the site of desired drug action. Thus, for example, lenticular wafers impregnated with the drug can be used for treatment of the eye, impregnated tampons as vaginal contraceptives. Throughout, the aim is to use the minimum of drug substance in the best possible place.

Marketing

Much adverse publicity has been engendered by the money spent by pharmaceutical companies in promoting their products. Why this should be so is difficult to explain, unless it results from the nature of the market itself. Maintenance of health and cure of disease are highly emotive, and the customers of the industry are both limited in number and articulate. Moreover when general practioners numbering perhaps fewer than 30 000 are courted by almost 100 companies promotion seems overly intense. In fact, the actual percentage of total sales spent on promotion in the drug industry is much less than in the soap industry. Even so, the highly technical character of drug promotion depends in major part on medical representatives interviewing one doctor at a time, economically a most expensive way of presenting products. But present them they must, since without such promotion few drugs stand a chance of rapid acceptance and it is speed that counts. The reason for this is that innovation takes a lot of cash and this cash has to be recovered as quickly as possible. The market situation of a new drug changes with time. In the U.K. for example the patent life of a compound (from the date of filing) is only sixteen years, and eight of these are spent before ever the material reaches the market. Moreover, the growing complexity both of discovery and its implementation will make an even larger proportion of the patent's life 'dead time'. If this trend continues, medicinal research will concentrate on those big issues which allow investment capital to be quickly recouped. Research into marginal ailments and tropical medicine, though still needed, will diminish.

5. The future of the industry

J. F. CAVALLA

IT is axiomatic that the pharmaceutical industry cannot continue to grow at its present rate for many more years. Even now there are signs of a plateau in the growth curve; within the next ten years, signs of stability will be clear. Yet there is still much to do and the advances which have been made so far are themselves responsible for generating problems requiring solution in the years to come.

The long-standing problems include cancer and viral diseases, hardly touched as yet by the industry. The new problems include the so-called geriatric diseases: heart disease, stroke, rheumatoid arthritis, and mental degeneration. It is because many of the killer diseases, especially the bacterial infections, are now under control, that people who in earlier days would have died (e.g. from pneumonia) now live on and suffer from geriatric disease. As has been discussed earlier, the first step towards a cure of these and other diseases lies in establishing an animal model for study. Geriatric disease is difficult to study in animals and much knowledge is still to be acquired; obviously the solution will not prove simple.

But this aspect of disease concerns mainly the western or industrialized world where the rate of population increase is declining. The other two thirds, soon to be three quarters, of the world population have no geriatric disease problems. There, the problem is how to live on into one's thirties and forties, in the midst of tropical infection, starvation, and lack of the most rudimentary sanitation.

Without doubt the most dramatic thing the West has done for the East is to provide cheap and effective insecticides and thereby reduce the spread of insect-borne disease. As a result, the numbers of babies born and surviving in these poor areas have increased at a time when no provision was made for their reception. The industry has provided remedies for most of the tropical diseases but almost all these are too expensive for the mass of the population inhabiting the tropics. It has also provided means of birth control which can be used by the least sophisticated populations, but their use has not yet become acceptable in the areas most in need of them. It is conceivable that, with the present low dosages required, oral contraceptives could be used as dietary additives, but such procedures offend against many ethnic and religious taboos. Thus it will be many years before equilibrium is reached between population increase on the one hand and a better healthier life on the other.

The pharmaceutical industry is a product of the success of industrial society and while all nations receive modern medicines the proportion of the population enjoying them is related directly to the affluence (i.e. gross national product per head of population) of the country concerned. In the developed countries, the demands made upon the pharmaceutical industry are for ever better medicines of ever greater safety. The ratio of compounds marketed to those made in the laboratories is continually changing adversely for the industry. What had been a ratio of 1 in 3000 in 1950 has become a ratio of 1 in 8000 by 1970. In part this is due to the drug regulatory authorities which were set up with great rapidity in most of the industrialized countries following the thalidomide disaster.

Since the whole existence of these commissions depends on not permitting another tragedy to occur, their requirements concerning the safety of new drugs are predictably severe. Such demands are to be welcomed in both maintaining and improving standards of medicine production and control. Nevertheless, there are voices within the industry which state that to demand excessive standards of freedom from toxic effects in a new medicine may well result in the stultification of drug research and ultimately of progress. Thus Lord Todd, a Nobel Laureate, has said 'I sometimes wonder if public authorities are not becoming obsessed by the word "safety" in connection with drugs. The demands made by, for example, the Food and Drug Administration in the United States for "safety" in the case of any new drug a pharmaceutical company wishes to introduce are becoming so fantastic in their financial implications as to render the development of a new drug well nigh impossible for any but a very large organization. This is not to say that safety is unimportant—it is extremely important. But what exactly do we mean by "safety"? No progress is ever made without risk and in the case of a new drug we are concerned with a balance of risk. To my mind one must ask oneself whether on balance benefit outweighs risk in this as in all questions related to human welfare'.

Medicines today are the one area where risks are seldom taken except in life-threatening conditions. Cosmetic surgery with all its attendant hazards is uncontrolled as are cosmetics themselves. But drugs now require a standard of perfection that results, at best, in delaying patient access for many years and, at worst, by preventing their very emergence into the doctors' hands. No doubt, as the years pass, a realistic equilibrium will be attained between drug safety and the patient's welfare but in the meantime good drugs may have been lost.

What of the more distant future? What will be the research programme for 1984? It may lie in the area of 'quality enhancing' drugs. Just as now we have compounds available for alleviating clinical depression and anxiety, so we may eventually have recourse to safer, non-addicting, non-debilitating compounds to replace our present alcohol and nicotine as chemical crutches that for some make the world a little happier.

True cosmetic chemicals, not dyes and emollient creams, are also not beyond the skill of the pharmaceutical industry. There might be compounds capable of staying the skin changes occurring at the menopause or the loss of natural hair growth in men with advancing years; prophylactic compounds to protect man from himself, a combined contraceptive and gonococcal agent for example, or from his environment with a memory improver. These are quite feasible innovations. Drugs are being discussed that may determine the sex of a child with an exactness now unthinkable: the effect on society could be immense. Anorexic agents, or appetite suppressants, may well be tailored to allow all weight reducing to be the simplest of procedures; this could avoid strain on the cardiovascular system and much needless stress.

At present, much of the above is only of marginal interest to scientists in the pharmaceutical industry. Extrapolation of the advances made in the last twenty-five years into the next fifty, however, (even allowing for the plateau now appearing) indicates that we shall have cures for heart disease, rheumatoid arthritis, and cancer, and capacity over and above that to allow for new developments. Work in such an environment will demand ability, invention, and determination; its rewards will be a freer and happier life for the community.

6. The agrochemicals industry today
D. PRICE JONES

Introduction

'AGROCHEMICALS' is the customer-oriented name applied to that sector of the chemical industry serving agriculture. It supplies materials for the control of various pests, plant diseases, and weeds, or for the promotion or regulation of plant growth. Agrochemical activities have close technological links with other chemical activities and usually form one part of a larger chemical organization. The closest affinity of agrochemicals, however, is with pharmaceuticals: both regulate living systems, but while the pharmaceutical industry is preoccupied with the health of individuals (even when many individuals are affected), the agrochemical industry thinks in terms of populations. This difference is important.

Many 'agrochemicals' are used outside agriculture. Thus, certain insecticides control insect vectors of human disease, and help to meet the objectives of the pharmaceuticals industry; some fungicides and bactericides preserve domestic or industrial materials, for example, paper pulp in pulp mills. Material finding such uses may represent spin-off from research aimed primarily at agriculture.

The two main objectives of the industry, pest control and regulation of plant growth, may be either separated or integrated commercial activities. Growth-promoting chemicals include fertilizers that have some of the characteristics of bulk commodity chemicals. The fertilizer industry is now mature: after past extensive research, development and optimization, it now offers relatively little scope for radical innovation and sophistication. In contrast, pesticides are newer. In their variety, sophistication, specialized use, and value per ton and in their exacting specifications, usually set by regulating authorities, they may be regarded as fine chemicals. Bulk chemicals are expensive to transport (when compared with the value of the product); for fine chemicals the transport cost is relatively small. Thus the fertilizer business tends to be regional while the pesticide (like the pharmaceutical) business is truly world-wide. In their agricultural effects, fertilizers and pesticides interact strongly: in any new agricultural project, fertilizer and pesticide developments tend to proceed concurrently. Plant growth regulants, long grouped with pesticides as 'crop protection products', are much less well known than are pesticides. In use since the 1930s, they have become significant commercially in the past decade. Their biological affinities lie with fertilizers—they modify the growth that fertilizers promote. Technologically, they resemble pesticides—they are small-volume chemicals, formulated and used in equipment designed primarily for pesticides.

The earliest use of pesticides by man is lost in antiquity. The development of settled agriculture demanded the safeguarding of food supplies in the field and in store. As man increased his range of foods so he encountered a wider range of pests. In settled agriculture and in later urbanization, the commensals—rats, mice, for example—became major pests. As goods acquired economic significance, so the pests that destroyed them developed an economic significance of their own. In the eighteenth and nineteenth centuries, therefore, pest control gained importance. By the end of the Second World War, the stage was set for a major revolution in pest control that affected the very structure of civilization. Triggered by the discovery of the insecticides (DDT and BHC, p. 52) and the selective weed-killers (MCPA and 2,4-D, p. 61), much industrial effort was put into the further synthesis and biological evaluation of organic compounds. A wide range of products resulted, covering almost every conceivable use for pesticides and achieving levels of control undreamt of in the 1930s. Furthermore, new, cheaper, and more effective application techniques were devised. At the same time, the biology of pests and the economics of pest damage were more closely studied.

The social effects of agrochemicals have been immense. The pest control system, as a social system, is not only complex, with its many interdependent parts; it is also remarkably well-adjusted to the needs of society. These needs include good health and adequate nutrition. The contribution of pesticides to health may never be fully quantified but the United Nations World Health Organization affirms that DDT and other insecticides have played a major role in the saving of millions of human lives that would otherwise have been lost to malaria. There are many other lethal, and also socially significant debilitating diseases, to the control of which pesticides make massive contributions. Crop losses due to pests and diseases are also difficult to quantify but the United Nations Food and Agriculture Organization puts such losses at about $200 000 million annually. These losses occur mainly in the developing countries where pest control measures are not as widely or effectively used as in the developed countries. By any criterion, pest control has a major role to play in society. The contribution of fertilizers to society is inextricably integrated with the improvements in agriculture over the past fifty years. The significance of synthetic plant nutrients in increasing the world's food supply is underlined by the FAO 'Freedom from Hunger' campaign, the basis of which is the encouragement of the use of fertilizers.

The agrochemical industry interacts with society not only in safeguarding health and food supplies, but also in its effects on the environment. The views of society on disturbances to the environment are evolving. Changes that are acceptable to sick and hungry people, may not be acceptable to the healthy and affluent. In this respect the industry's markets differ geographically. How agrochemicals interact with the environment and how new technology is developed in response to changes in the wishes of society are topics discussed later. In this introduction it is worth emphasizing that change is continuous: some changes are forced on the industry, but many arise from within the industry itself as its highly responsible personnel—many of them scientists—wish to ensure that the industry and the company of which they are part meet society's current and foreseeable demands.

Research on agrochemicals

Chemical innovation in pesticides and plant growth regulators differs greatly from that in fertilizers, and there are some, but much smaller, differences between pesticides and growth regulators. It is better, therefore, to discuss innovation in agrochemicals separately for each of these groups.

Pesticides

The original pesticides were discovered by chance or the result of operations in which random selection played a major part. Even today most original discoveries arise in this way. Once an original discovery is made, random effects are reduced by managerial action and control. Further research and development, by almost standard methods, continue until either a useful product emerges, or it becomes clear that no useful product is likely to emerge. As long as the mechanism of a biological effect is unknown, new activity can be discovered only by random synthesis and testing. Empirical chemical modification of a substance which has some activity but is unsuitable for use provides another route to new products. The pesticide industry closely resembles the pharmaceutical industry in this; unfortunately few mechanistic speculations are well established, even though many useful pesticides are known.

One difficulty the industry faces in trying to produce novel better pesticides (where 'better' may be defined in specific ways, e.g. value for money or fewer side-effects) is to decide which biological effects are significant. Once this is decided, a large number of compounds (often randomly selected, often based on intuition) is submitted to the appropriate screening tests. If biological activity is found, this is regarded as an original discovery. Related compounds are synthesized and tested; as in pharmaceuticals, empirical concepts relating structure with activity are developed to guide further work.

When only one specific effect is sought, activity can sometimes be correlated with a number of chemical or physical parameters. Where multiple effects are involved, the final decision is reached only by subjective judgement. Nevertheless, in this manner some understanding is reached of the way changes in the molecule influence the biological effect. With more detailed studies, it may be possible to relate the physicochemical properties of the molecule to the ultimate biological activity. But there are many steps in this process: for example, a stomach insecticide, applied to a leaf surface for the control of a caterpillar, must possess certain qualities, some of which relate to the active material itself and some to the formulation. For the present, only the active material is considered. The insecticide, preferably with a low mammalian toxicity, must have an adequate shelf-life at temperatures to which it is likely to be exposed. Once on the leaf, it must be reasonably resistant to weathering and to u.v. light (formulation may assist). It must not damage the leaf (must not be phytotoxic), nor repel the caterpillar. Once ingested, it must resist degradation by insect and plant enzymes and be stable under conditions of changing pH. It must have the right physico-chemical properties to penetrate the gut, but retain resistance to enzymes. In the insect's circulatory system, it must resist detoxication mechanisms. If it happens to be a nerve poison, it may still have to penetrate the nerve sheath before arriving at the site of action. One chemical is therefore required to possess a very special set of physico-chemical properties (for example, suitable partition coefficients, and resistance to hydrolysis and oxidation) enabling it to traverse a series of biological situations before it can eventually perform its allotted task. Many potential biological blocks exist between the point of application and the site of action, each offering its own set of problems and hazards for the invading chemical. Chemicals applied to the soil may have to resist leaching, microbial attack, chemical degradation or inactivation through adsorption on soil particles, and still pass through all biological and physical barriers in the pest organism on the way to the site of action.

Partly because of this complexity, we lack the knowledge to design chemicals that by-pass these barriers and to relate physico-chemical properties directly to the ultimate biological activity. Thus physico-chemical experiments such as *in vitro* tests on enzyme systems bear little relation to reality. In this respect pesticide research, particularly with plants, differs from pharmaceutical research where *in vitro* experimentation is widely used to reduce the amount of animal experimentation, which can be both distasteful and expensive. In crop protection research, it is usually as easy to screen on the whole organism, where the effect may be directly observed. There is the added bonus that other important direct or side-effects may be simultaneously observed. Pesticide research is concerned with a set of biological effects, of which the ultimate control of the pest (weed, disease, or whatever) is but one. It may not even be the most important effect: for instance, to put it mildly, it is more important not to kill the spray operator or the crop, than to succeed in killing the pest. To be successful in this task, it is necessary to accumulate a wide range of facts and order them somehow, so that the knowledge means something and is useful for prediction.

Plant growth regulators

Problems in innovation sought for plant growth regulators are very similar to those for pesticides. The one notable difference—mainly a matter of emphasis—is the crucial importance of the physiology of plant growth and development. Our rapidly increasing knowledge of the biochemical systems involved has begun to influence chemical synthesis. This may be seen, for instance, in the concentration of effort on phenoxycarboxylic acids in the 1940s and more recently—and more significantly—on so-called 'ethylene generators' (compounds capable of releasing ethylene within the plant). However, even this research still has a large random component. Biological screens, based mainly on whole plants, are employed to throw up leads. The subsequent steps are strictly analogous with those adopted in pesticide research.

Fertilizers

Plant nutrients (p. 66) are already well known. Current research thus differs from that on pesticides and plant growth regulants. It is less concerned with novel chemicals than with cheaper production methods, with physical conditioning to allow better handling, storage and application, and with ensuring adequate and timely availability of nutrients to given crops under a variety of conditions. Work on nitrogenous fertilizers is influenced by:

(1) The need for high concentration (which reduces transport costs); urea is a material 'designed' for this purpose.
(2) The need for a form in which other nutrients, especially phosphates, can be included (e.g. ammonium phosphate).
(3) The increasing demand for slow release, thus diminishing plant injury, extending availability to the crop, and reducing drainage into rivers and lakes, with the associated risk of eutrophication (p. 80).

None of these tasks resembles the designing of a new pesticide. Even the search for new slow-release materials is in a quite different category. Innovative work on new production processes continues; and research into biological fixation systems (as found in blue-green algae or legumes) may provide new scientific background to this large industry.

7. Specific biological effects

D. PRICE JONES

Control of pests—general

THE term pest is normally applied to an organism that interferes in some way with man. It may affect his health or convenience, his animals or his goods, or imperil his food supply. As man has dispersed over the earth and extended his activities, he has greatly increased the number and variety of organisms that interact with him, including the number of pests. Man's cultural evolution has reached a stage where he demands a high standard of health, nutrition, and comfort, including freedom from petty annoyances and protection of his goods from pests.

The variety of pests is very great indeed. They include viruses and numerous representatives of almost every major group of plants and animals. Those noxious organisms, e.g. certain bacteria and viruses, causing disease in man and animals come within the purview of the pharmaceutical industry and are discussed earlier in this book. Other broad categories of pests, for example crop pests, household pests, or pests of stored products, generate quite different problems and require different solutions. Control of pests is not easily compartmentalized: pesticides interact strongly with the environment, including other organisms and agencies regulating pest abundance. The full implications of this will emerge as the discussion develops.

As pests exist in great variety and affect man in many different ways, so the chemical agents used for the control of these pests show endless variation. Many can be grouped in acceptable, if rather general, chemical classes, such as organophosphates, carbamates, triazines, substituted ureas or pyrimidines; others, such as rotenone 75 are apparently isolated examples of pesticidal activity. A classification more commonly adopted by biologists, agronomists and others involved in field applications, designates the type of pest against which the agent is employed. Thus, insecticides are used for the control of insects, rodenticides for rats and mice, nematicides for eelworms, molluscicides for slugs and snails, acaricides for plant feeding mites, herbicides for weeds, fungicides for fungus diseases and bactericides for bacterial diseases. The term 'biocide' implies a destroyer of life, but its use is increasingly restricted to those materials used for the control of biological agents (commonly bacteria and fungi) responsible for decay or other deterioration of man's worldly goods. Other terms, such as 'algicide' (controls algae, e.g. the green scum often found on stagnant water), or 'viricide' (controls viruses), are coming into general use but are not yet completely assimilated.

Control of insect pests

Insect species account for roughly two-thirds of all known species of animals and plants; and insect pests, again as species, are far more numerous

Specific biological effects 51

than other pests (the Commonwealth Institute of Entomology has recorded some 70 000 insect species). Insect pests are mostly visible to the naked eye and many have, as a consequence, been known from ancient times, whereas bacteria and fungi had to await the invention of the microscope for proper recognition. It is not surprising, therefore, that insecticides have been in use much longer than other classes of pesticide (rodenticides excepted); currently they are more numerous than other types of pesticides.

Plants have evolved under insect attack for some 300–400 million years and have developed a variety of chemical mechanisms to protect themselves against insects. These include deterrence, repellency, and insecticidal action. Some of these naturally-occurring insecticides have been used by man since time immemorial without any knowledge, until recently, of their chemistry. A few have survived to the present day, the best-known representatives being derris (rotenone **75**) from *Derris elliptica*, nicotine **76** from tobacco, and pyrethrum **77** from the flowers of *Pyrethrum cineraefolium*. It is a popular fallacy that natural insecticides of plant origin are safe in use. Nicotine is

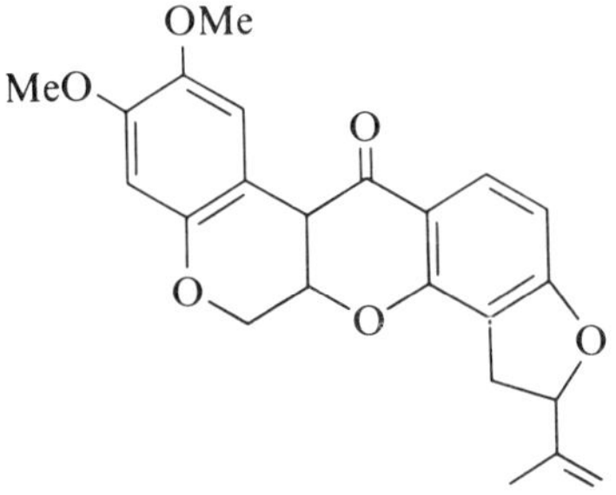

75 Rotenone

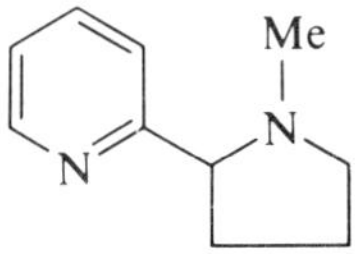

76 Nicotine

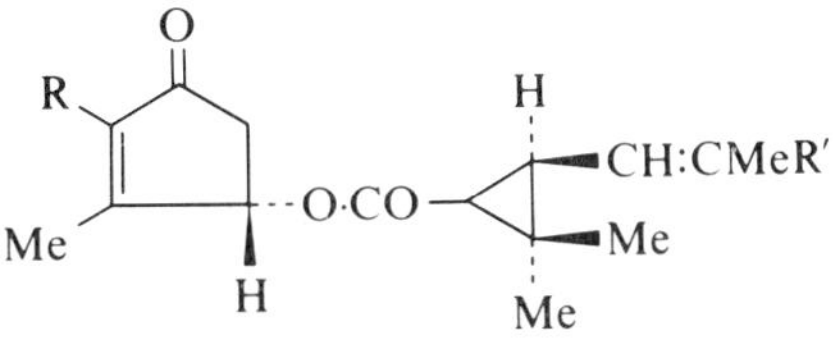

77 Pyrethrins (R = $CH_2 \cdot CH:CH \cdot CH:CH_2$)
   a Pyrethrin I, R' = Me
   b Pyrethrin II, R' = CO_2Me

extremely poisonous and derris is a highly effective fish poison (its main traditional use). However, pyrethrins are relatively harmless to most other forms of life.

Insecticides in use before 1940 were largely inorganic compounds of lead, mercury, or arsenic. Lead arsenate used for the control of the apple codling moth led to resistance in the pest and to increasing concentrations in orchard soils in parts of the U.S.A., with subsequent injury to trees. Arsenical poisoning through beer in the U.K. was traced to lead arsenate used on hops. These are among the early records of side-effects of pesticides.

Synthetic insecticides form three broad classes of compounds; organochlorines, organophosphates and carbamates. Initially, the discovery of DDT **78** in 1939 and of BHC **79** in 1942 led to the realization that organochlorines in general may be insecticidal. A spate of organic syntheses followed, along two main lines: first, preparation of analogues of DDT and BHC and of other chlorides; second, near random chlorination of hydrocarbons. Out of this tremendous effort, directed into chemical synthesis and biological evaluation, came a rash of related compounds loosely known as 'organochlorines' or 'chlorinated hydrocarbons'. They undoubtedly formed the spearhead of the massive development of pesticides in the 1950s and 1960s and have also been the prime cause of the equally massive reaction in the popular press against pesticides in general and insecticides in particular. The story is instructive. DDT was first used by the Allied armies during the

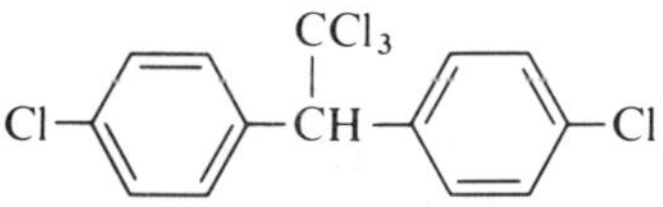

78 DDT 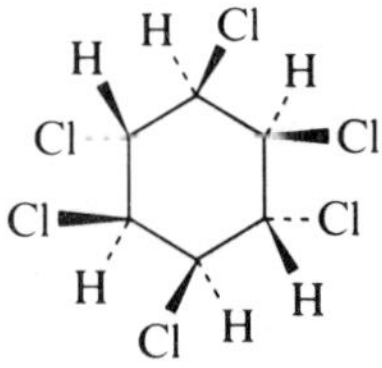   **79** γ-BHC (Lindane) 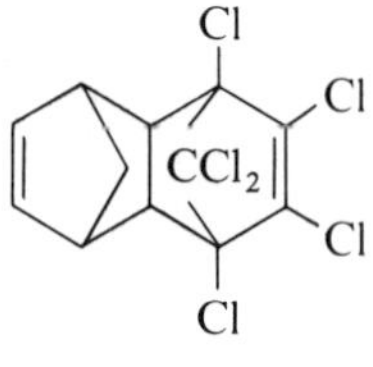   **80** Aldrin

Second World War to control the mosquito vectors of malaria and the louse vector of typhus. In post-war years, it was rapidly developed as an agricultural insecticide. The properties that were most exploited were its relatively low mammalian toxicity, the broad spectrum of activity, the remarkable persistence, and the high toxicity to insects. Its persistence, demanding less accuracy in timing and often fewer applications, made DDT easy to use. The wide spectrum made it useful against a wide range of insects. It soon became a large tonnage insecticide, easily outclassing all other pesticides. The outstanding contribution of DDT was in the control, and even eradication, of malaria in many parts of the world.

Other organochlorines followed DDT. The cyclodienes, for example aldrin **80**, represent developments from one original discovery. Although they have contributed greatly to the control, particularly of mosquitoes and cotton pests, they suffer from many of the disadvantages of DDT; most important, they are generally more toxic than is DDT. They were responsible for large-scale deaths among grain-eating birds in Britain in the 1950s and there is

Specific biological effects

circumstantial evidence to suggest that they contributed to deaths among predatory birds. The organochlorines were being phased out in many developed countries in the 1960s and early 70s, largely because of their persistence and their entry into food chains. The reason that the use of these valuable insecticides can be discontinued is the availability of other new, less persistent chemicals, often with narrower spectra of activity. Discoveries in Germany in the late 1930s and during the Second World War drew attention to the vast potential of certain organophosphates. The first three compounds pointed the way to future developments: tetraethyl pyrophosphate (TEPP) **81** provided a quick clean-up of aphids (greenfly) on lettuce. Because the toxic residues disappeared rapidly, early marketing was possible. Parathion **82** proved to be a powerful broad-spectrum insecticide with moderate persistence; and schradan **83** was the first systemic insecticide. For many years, the main restriction on the development of organophosphates was their high mammalian toxicity. The restriction was removed by the discovery of malathion **84** which is rapidly detoxicated in the mammal.

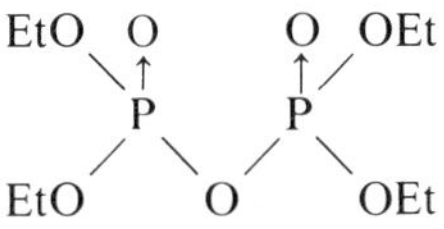

81 TEPP

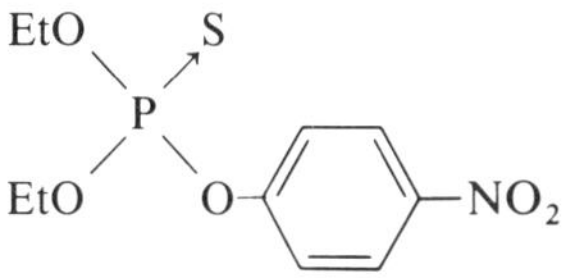

82 Parathion

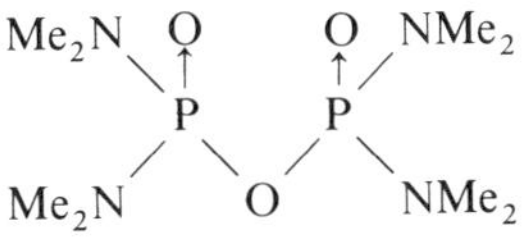

83 Schradan

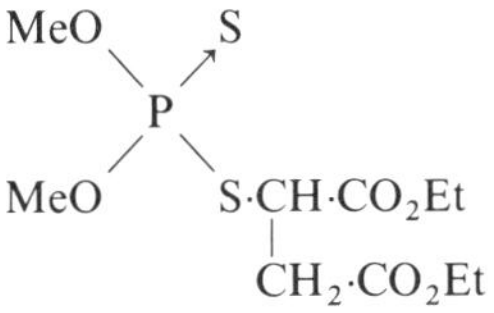

84 Malathion

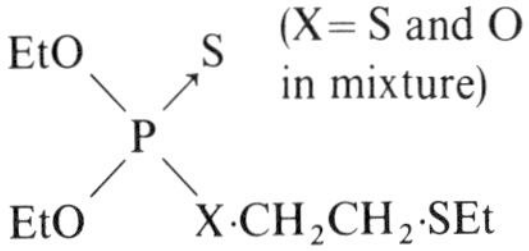

85 Demeton (X = S and O in mixture)

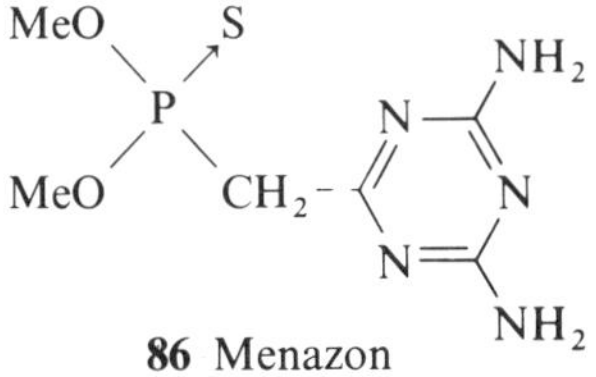

86 Menazon

The organophosphates penetrate into plants and then move through the vascular system to distant parts of the plants. This 'systemic' movement avoids the need for treating specifically an infested area. Absorption can occur through the roots, stem, leaves and (sometimes) fruit, but most commonly through the roots and leaves. Movement is generally from the roots towards the leaves, or towards the tips within the leaves. Some exceptional (and for this reason very valuable) materials, e.g. demeton **85**, can move in the reverse direction. This ability to move systemically can be exploited both to improve the kill of the target pest and to minimize the effect on other insects. Some organophosphates, e.g. menazon **86**, are highly selective in their insecticidal activity and do not damage beneficial insects: they point the way to a more general development of selectivity. Organophosphates affect nervous transmission in both mammals and insects, largely through the inhibition of acetylcholinesterase, the normal role of which is to remove acetylcholine (produced during nervous transmission). Organophosphates have caused many deaths among workers involved in their application, usually through neglect of precautions.

The third major group of insecticides, the carbamates, had been known for some years. The discovery that boosted the development of this group was that of carbaryl **87** introduced in 1956. A broad-spectrum insecticide with slight systemic properties and a low mammalian toxicity, carbaryl has been used against many insect pests, especially on fruit, vegetables, and cotton. In recent years, very many additional carbamate insecticides have been marketed, with (collectively) a gamut of properties similar to that of the organophosphates, even in their acetylcholinesterase inhibition. Some have short, others moderate persistence; some display systemic activity; at least one, pirimicarb **88**, is highly selective against aphids, and relatively harmless to beneficial species.

The story of insecticides so far thus tells of successive evolutionary steps from non-selective toxic inorganics to the first really effective insecticides of immeasurable benefit to man, DDT and BHC. Attempts at improvement and greater sophistication have provided advances and set-backs: recognition of environmental damage and discontinuation of use of the most persistent toxic organochlorines; sophistication of spraying techniques to prevent accidents in the spraying of the 'nerve-gas' organophosphates; the recognition of systemic insecticidal action; and most recently the welcome discovery of selectivity in carbamates.

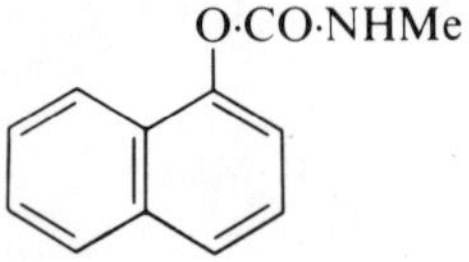

87 Carbaryl

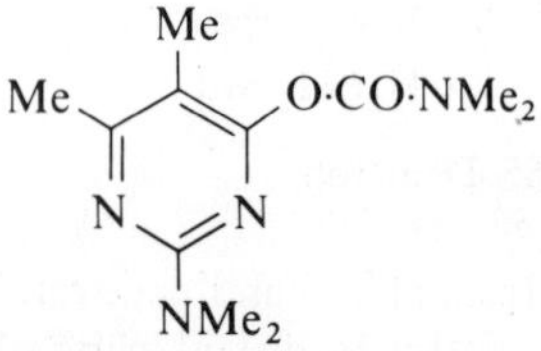

88 Pirimicarb

Specific biological effects

Rodenticides

Rodents include many species classifiable as pests: the brown rat is the main pest in temperate regions and the black rat—its fleas transmit bubonic plague—in warmer climates. Rats cause enormous losses to stored products, especially foodstuffs. The house mouse may not cause as much economic damage, but it is cordially disliked. Field mice sometimes cause local devastation in arable crops or young forest plantations, particularly in warm temperate regions.

Rats and mice have been commensals of man ('eating at the same table') for thousands of years and impeded the development of a settled agriculture by interfering with stocks of stored grain. Rodenticides were in use in ancient Babylon and in Egypt. In recent times, materials such as elemental yellow phosphorus, red squill, and strychnine have been used. Red squill, an extract of *Urginea maritima* containing the glycoside scilliroside, is reputedly safe for man because of its emetic properties, but there could be significant—and fatal—exceptions: rats cannot vomit and are therefore susceptible. The use of red squill and strychnine is prohibited on humane grounds in certain countries. Other acute poisons highly toxic to rats include alphachloralose, fluoroacetamide, zinc phosphide, arsenious oxide, and thallium sulphate, but their use, too, causes difficulties.

In recent years, various anticoagulants, e.g. warfarin **89**, have been widely and successfully used against both rats and mice. These are slow acting (chronic) poisons, causing haemorrhage, ultimately fatal. Although anticoagulants are not markedly selective, it is relatively easy to ensure that domestic animals do not ingest them repeatedly, and these animals are not normally adversely affected. Moreover, in cases of accidental ingestion, treatment with vitamin K_1 is an effective counter-measure. These rodenticides have been in use for some years and resistance has now appeared in some areas; this threatens their utility and is likely to prove costly, but challenges the chemist to further innovation.

89 Warfarin

Control of plant diseases

Crop diseases were recognized in ancient times. Thus the Babylonians had a magic formula for dispelling cereal smut. The true pathogenic nature of these diseases was not, however, established until the mid-nineteenth century. Epidemics of 'late blight' then ravaged potato crops in Northern Europe and culminated in the Irish famines of 1844 and 1845. They prompted scientific investigation of plant diseases. To de Bary goes most of the credit for demonstrating, during the 1850s, the pathogenicity of various fungi, and for stimulating the understanding and rational control of the important pathogenic species.

The potato blight is an outstanding example of the devastation and social consequences that can follow epidemics of plant disease but there are many others. Towards the end of the nineteenth century the coffee crop in some regions was devastated by coffee leaf rust; in Ceylon, plantation coffee had to be grubbed and replaced by tea. This same fungus now threatens the Brazilian coffee crop. Fungal diseases which attack other tropical crops, include blackpod of cocoa, sigatoka of bananas, blister blight of tea, vascular wilt of palms, and blast disease of rice—all of great economic importance. Bacterial blight of cotton is an example of an important bacterial plant disease. A fungus growing on groundnuts stored under damp conditions produces the remarkably toxic aflotoxin responsible for the deaths of many thousand turkeys in Britain in the 1950s, and believed to be responsible for the high incidence of certain stomach diseases in man in various parts of the world. In temperate regions, various fungus diseases of cereals are attracting increasing attention as cereal farming becomes progressively intensified.

Prevost, as early as 1807, showed not only that bunt disease of wheat was caused by a fungus but that this fungus could be killed by copper sulphate. Acceptance of these findings was slow. It was not until the end of the nineteenth century that copper sulphate alone, or in combination with lime or sodium carbonate (Bordeaux and Burgundy mixture, respectively), became established as a commercial fungicide. Such mixtures controlled downy mildew on vines and this provided the stimulus for the invention and production of knapsack and mobile sprayers which, in turn, did much to further the development of commercial spraying. Other materials in use at the turn of the century were sulphur, lime/sulphur mixture (calcium polysulphides of indeterminate composition) and formaldehyde. All these were protective rather than eradicant: they had to be applied before the disease developed and preferably before infection occurred. Thus, they were often applied unnecessarily. Such happenings served to highlight the phytotoxicity (toxicity to plants) of these materials and to encourage the search for new selective fungicides.

Copper compounds remain in commercial use on a diminishing scale. Both copper oxychloride and cuprous oxide underwent progressive sophistication in manufacture and formulation in the 1950s and emerged as valuable fungicides, against downy mildew and leaf spot, with diminished phytotoxicity.

Specific biological effects

Inorganic mercurials although widely used for the steep treatment of seed, corms and bulbs, were too phytotoxic for use as fungicides. The search for less injurious materials led to organomercurials of the type R–Hg–X, where R is an alkyl or aryl radical and X an inorganic or organic radical. Several have achieved widespread use as seed dressings. However, alkylmercury, particularly methylmercury compounds, produce irreversible effects on the mammalian central nervous system. The feathers of raptorial birds in Sweden showed a marked increase in mercury content associated with the introduction of methylmercury dicyandiamide **90** seed dressings. The use of mercurial fungicides is now under examination because of danger to birds and fish and indirectly to consumers of contaminated fish. The laboratory demonstration of the convertibility of phenylmercury into alkylmercury capable of re-entering environmental cycles arouses further misgivings.

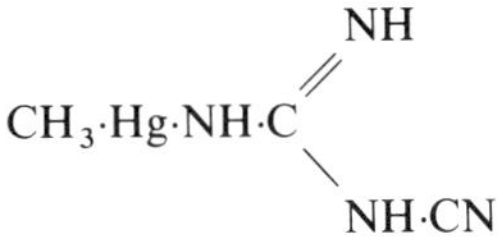

$$CH_3 \cdot Hg \cdot NH \cdot C \begin{array}{c} \diagup NH \\ \diagdown NH \cdot CN \end{array}$$

90

Other fungicides that have been in use for decades are metal complexes of dithiocarbamates. They have antecedents in the rubber industry, and were discovered in the U.S.A. in the early 1930s. Ferbam **91a** and ziram **91b** were among the first to be marketed. As others of the series appeared (e.g. **92**) they came to be known collectively and elegantly as the 'educated sulphurs with the biblical names'. They replaced mixtures of sulphur or lime/sulphur which had always posed problems of phytotoxicity and erratic performance.

$[R_2N \cdot CS \cdot S]_n \cdot metal$

91a Ziram, R = CH_3, metal = Zn
91b Ferbam, R = CH_3, metal = Fe

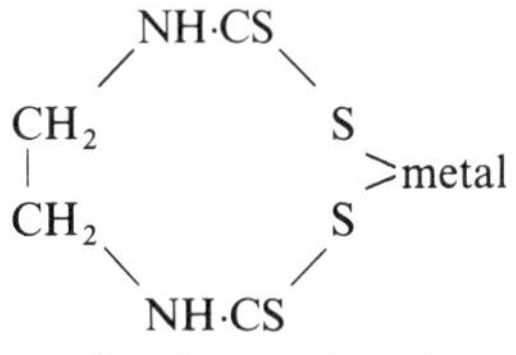

$$\begin{array}{c} NH \cdot CS \diagdown \\ CH_2 \qquad\qquad S \\ | \qquad\qquad\qquad >metal \\ CH_2 \qquad\qquad S \\ \diagdown NH \cdot CS \diagup \end{array}$$

92a Zineb, metal = Zn
92b Maneb, metal = Mn

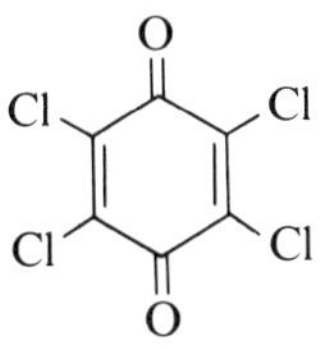

93 Chloranil

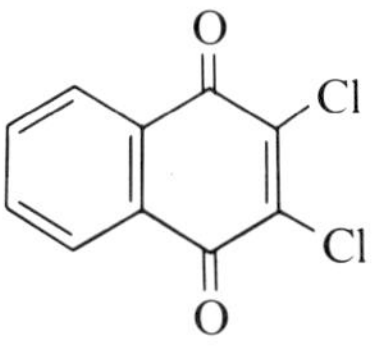

94 Dichlone

The dithiocarbamates were much less phytotoxic, and generally more potent and more consistent in their biological activity. They have found major and world-wide use as foliage fungicides. Zineb **92a**, in particular, has been used extensively on vegetables and also for the control of late blight on potato. It is effective against downy mildew on vines but the essential conservatism of the wine industry long resisted its replacement of copper.

The number of metal-free fungicides in use is growing. Among the older ones, the quinone fungicides, discovered in the 1930s, have not been exploited as foliar fungicides, possibly because of their sensitivity to u.v. light. Nevertheless, two of them, chloranil **93** and dichlone **94**, achieved widespread commercialization as seed treatment materials for corn (maize), beans of various kinds, and vegetables.

Aromatic dinitro-compounds, more particularly the dinitrophenols, have been known even longer. DNOC **95a** has been widely used as an insecticide and as a weedkiller. The search for less phytotoxic materials, and particularly for compounds with a significantly lower mammalian toxicity, led eventually to the marketing of several materials, of which the two closely related compounds dinocap **95b** and binapacryl **95c** have achieved commercial success. They have been used principally against powdery mildew diseases, especially on apples.

Captan **96**, introduced in the U.S.A. in 1949, marked a big improvement in fungicides. It has extremely low toxicity both to mammals and to plants. It was developed rapidly as a fungicide for foliage protection and proved valuable against apple scab (although displaced later by an analogue, folpet). Captan **96**, by reason of its low phytotoxicity, can also be used as a seed treatment and as a furrow treatment in cotton and other crops for the control of soil-borne diseases.

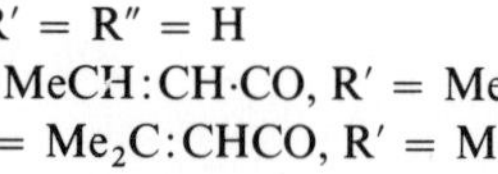

95a DNOC: $R = R' = R'' = H$
95b Dinocap, $R = \text{MeCH:CH·CO}$, $R' = \text{Me}$, $R'' = C_6H_{13}\text{-}n$
95c Binapacryl, $R = \text{Me}_2\text{C:CHCO}$, $R' = \text{Me}$, $R'' = \text{Et}$

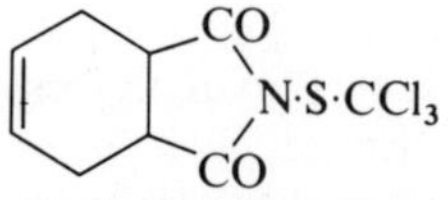

96 Captan

Specific biological effects

More recently a further major advance has seen the introduction of plant systemic fungicides. Their mode of action has attracted a great deal of attention, but they have many common biological features in addition to their systemicity. One of the earliest systemic fungicides, the antibiotic griseofulvin **97**, has only a limited outlet in horticulture. The others belong to five well-defined groups exemplified by carboxin, **98**, ethirimol **99**, benomyl **100**, thiophanate **101** and tridimorph **102**. These fungicides move upwards and outwards in plants: thus, watering the soil around the roots of certain plants ensures the arrival of the fungicides in, for example, leaf tissue where

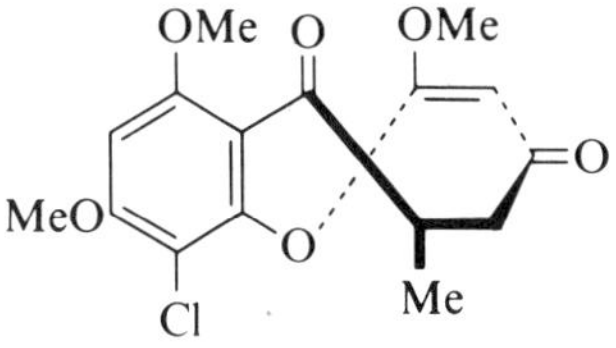

97 Griseofulvin

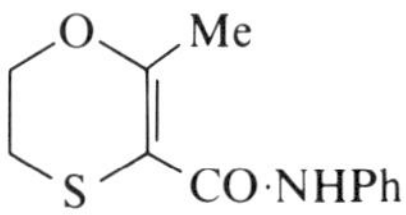

98 Carboxin

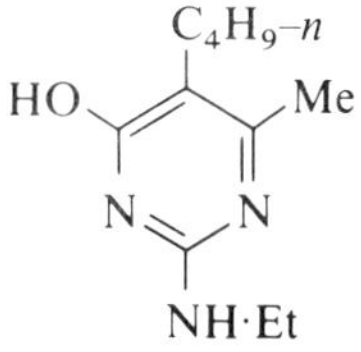

99 Ethirimol

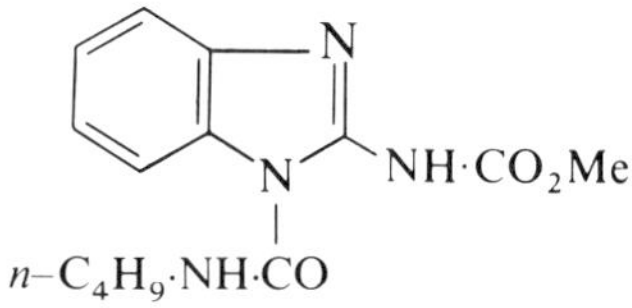

100 Benomyl

101 Thiophanate

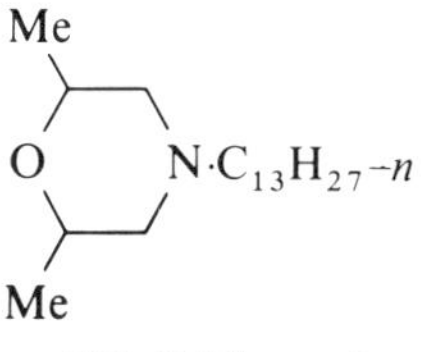

102 Tridimorph

their activity is best exploited. However, none of the existing materials is capable of moving downwards. In practice, systemic fungicides may be applied as a soil drench, as a seed dressing, or as a foliage spray. As a spray, the systemicity is not fully exploited but some local movement improves the distribution slightly. Also, systemicity is accompanied by some capacity for penetration, which in turn increases the scope for fungicidal activity.

All the current commercial products are active against powdery mildews, particularly against those on cucumber and barley. Ethirimol **99** has emerged as a valuable seed treatment for barley, while tridimorph **102** and benomyl **100** are used as foliage sprays. Benomyl with a rather broader spectrum than the others, is widely used against diseases of top fruit and vines.

Resistance problems, not hitherto commercially significant in fungicides, are now beginning to appear. For instance, after a long period of use of organomercurials as seed dressings, resistance to them has been observed in certain species of fungi. More significantly, resistance to some systemic fungicides has appeared in a relatively short time. This is particularly unwelcome at a time when the costs of developing and commercializing a new product are mounting rapidly.

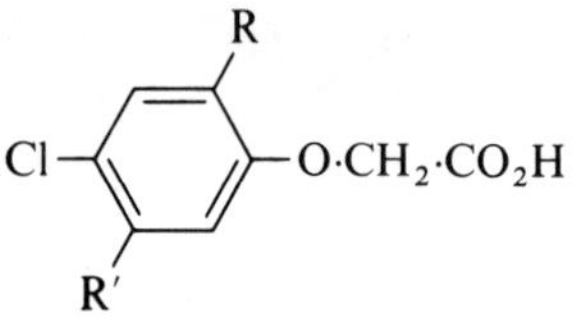

103a 2,4-D, R = Cl, R′ = H.
  b MCPA, R = Me, R′ = H
  c 2,4,5-T, R = R′ = Cl

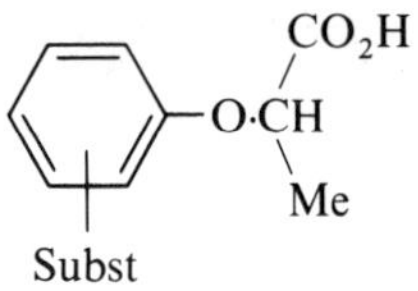

104

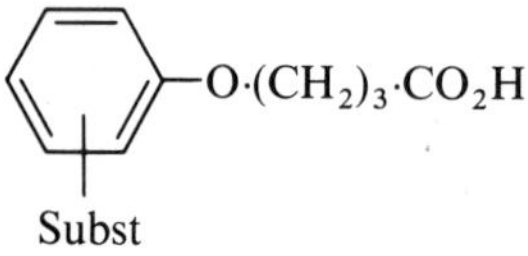

105

Control of weeds

Agricultural food production has always suffered from competition by weeds. Indeed, the spread and development of agriculture in Neolithic times across Europe can be traced by pollen counts in peat deposits—pollen counts of elm and other trees diminish as those of weeds increase. Control of weeds in primitive agriculture required manual labour and the development of agricultural systems and crop husbandry patterns such as shifting cultivation, ploughing, leaving land temporarily fallow, and row cropping. All these increased food production. Important steps forward accompanied the development of the drill and of horse-hoeing. The resultant improvement in labour productivity in part helped to fuel the first industrial revolution, by releasing workers previously employed on the land, who were then forced to seek employment in manufacturing industry.

Specific biological effects 61

Until the Second World War, improvement in agricultural weed control continued through innovation in equipment and husbandry techniques. Organic weedkillers like DNOC **95a** introduced in the 1930s, and well known inorganics like chlorates, arsenites, and sulphuric acid are total herbicides. They are useful, for example, for clearing paths, but had little impact on agriculture. All this changed in the 1940s: interest in plant auxins and in their synthetic mimics in the 1930s led to the almost simultaneous discovery in Britain and in the U.S.A. of the usefulness of the phenoxyacetic acids **103**. These compounds resemble the auxins in affecting the growth rates of plants and are, therefore, called 'hormone' weed-killers. Although they can be readily absorbed through the roots, in practice they enter mainly through the leaves or stems. They attack broad-leaved weeds selectively, leaving cereals unaffected. Dosage control and timing enhance the selectivity of the hormone weed-killers. In the late 1940s they began an agricultural revolution paving the way for mechanized, intensive cereal farming as practised today, barely twenty-five years later.

The success of these materials encouraged the synthesis and testing of many thousands of related compounds; the search was for more selective action against more weeds. Eventually, the 2-phenoxypropionics **104**, and 4-phenoxybutyrics **105** were found, controlling different spectra of weeds and filling gaps left by the original discoveries. The activity of the 4-phenoxybutyrics **105** depends on their ability, in many broad-leaved plants, to undergo β-oxidation, with *in situ* production of herbicidal phenoxyacetics **103**, but this β-oxidation does not occur in clover. Hence the butyrics **105** can be used in cereal crops undersown with grass-clover mixtures. There are many other examples of commercial herbicides whose selectivity offers advantages in specific situations. Apart from 'hormone' weed-killers, others are known with a different mode of action. Diazine, triazine, and thiocarbamate systems have been extensively investigated and used.

The potency of the hormone weed-killers is very high. They can be used in low volumes and this has led to major changes in spraying techniques with rapid increases in the areas treated.

The success of the phenoxyacetics in selectively removing broad-leaved weeds from cereals prompted a search for the reverse effect—the removal of graminaceous (grass or cereal) weeds from a broad-leaved crop. Propham (isopropyl *N*-phenylcarbamate) was the first to appear but met with only limited success; others followed. Perhaps the most successful in Britain is barban **106** used for the control of wild oats and blackgrass in cereals and other crops.

Since their discovery in the mid-1950s, the bipyridylium herbicides have attracted much attention. Diquat **107**, the first to be marketed, is active against broad-leaved plants and used mainly for potato haulm destruction as an aid to blight control and crop harvesting. Diquat is one of the very few herbicides that can be used for the control of aquatic weeds. Its use has to be restricted to small areas at any one time, so that dead vegetation does not deoxygenate the water too severely.

A second compound, paraquat **108** is effective against grasses as well as broad-leaved weeds. The bipyridylium herbicides act by trapping free radicals and thus interfering with photosynthesis. They are very rapidly inactivated by adsorption onto soil particles. Any crop seeds or unemerged crop seedlings are therefore unaffected. This has opened the way for the development of new crop husbandry techniques, summarized in the phrase 'minimum cultivation'. It is possible with paraquat to kill the residue of a crop and the attendant weed cover, and to drill almost immediately. Where the dead herbage is a hindrance, or the ground is too hard, minimum tillage, suited to the conditions, can precede drilling. Ploughless farming has advantages. The old stubble protects the soil surface and reduces erosion by wind and water. Avoiding the heavy tackle required for ploughing stiff clay soils, reduces the rate of deterioration of soil structure. Shortening the interval between harvest and replanting may enable three crops of rice to be grown where only two had been possible before. Thus, again, an agrochemical promises a major development in agricultural productivity.

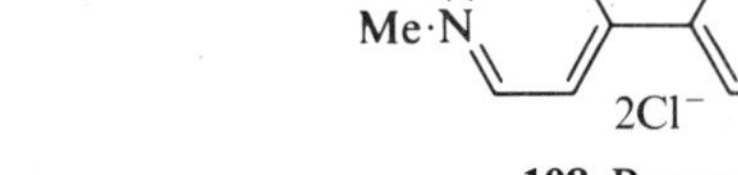

$$\text{C}_6\text{H}_5\text{-NH·CO}_2\text{·CH}_2\text{·C}\vcentcolon\text{C·CH}_2\text{Cl}$$

106 Barban

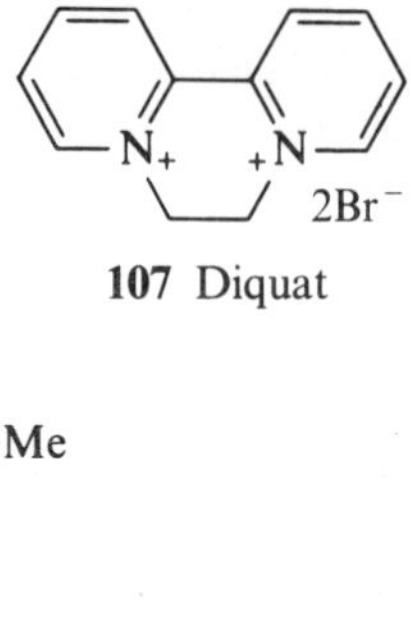

107 Diquat

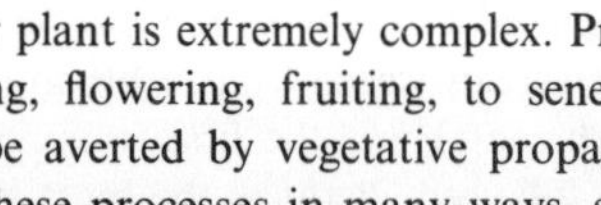

108 Paraquat

Regulation of plant growth

Growth in a higher plant is extremely complex. Processes range through germination, sprouting, flowering, fruiting, to senescence and death; or perhaps death may be averted by vegetative propagation. In agriculture, man interferes with these processes in many ways, sometimes unwittingly, sometimes deliberately. This interference may come from husbandry practices or—increasingly—through the use of growth regulating substances.

The study of plant growth regulating systems started, perhaps, with the observations in 1932 that acetylene and ethylene induced flowering in pineapple. In 1934 it was shown that indole-3-acetic acid **109** promoted growth of certain plant tissues and organs; within a few years it had been found in

Specific biological effects

some plants and its presence suspected in many others. The hunt was on: plant auxins had become a fashionable subject for research. Many naturally occurring chemicals are involved in plant growth regulation. On structural and physiological criteria, these are grouped as auxins, gibberellins, kinins and abscisins. Their effects overlap to some extent and they also interact strongly with one another.

Auxins, such as IAA **109** and a number of indole-3 derivatives of various carboxylic acids, influence many processes in plants but their most fundamental effect is induction of cell elongation. The apex of a plant bends as it grows towards the light—thus 'phototropism' results from the accumulation of auxin on the darker side of the growing tissue (or meristem). Likewise, the dominance of the growing point over adjacent lateral buds is due to its more vigorous production of auxin, which inhibits bud break. Among the other physiological processes affected by auxins are inhibition or stimulation (depending on circumstances) of growth and differentiation in various tissues and organs, leaf abscission, flowering, and sex expression in some plants, and fruit set and growth.

The gibberellins were derived originally from the fungus *Gibberella fujikuroi*, the agent causing bakanae disease in rice. Apart from the distortions of bakanae disease, one of the early distinguishing effects was the restoration of the tall habit to dwarf peas and beans. The gibberellins are believed to be important in mobilizing sugars in certain plant organs: some of their effects, e.g. hastening the germination of barley, may be due to this property. The growth of some fruit, e.g. pears, depends on the release of gibberellin by the seeds. Some thirty-two gibberellins are now known; they are polycyclic diterpenoids. The most widely investigated is GA_3 **110**. Activity is related to the lactone ring and the stereochemical configuration of the hydroxyl group on the A ring.

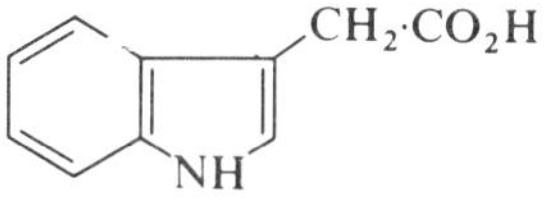

109 IAA

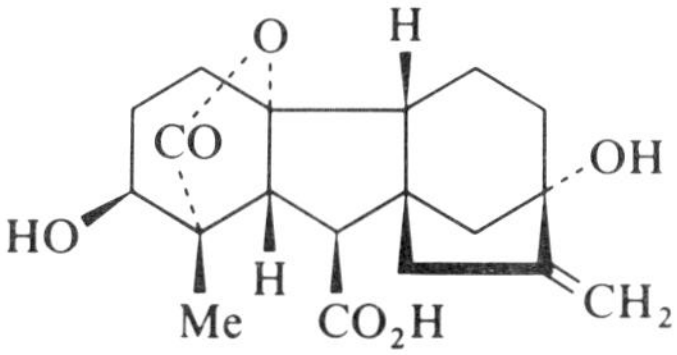

110 GA_3

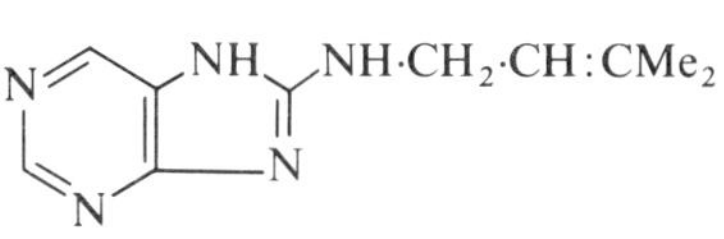

111

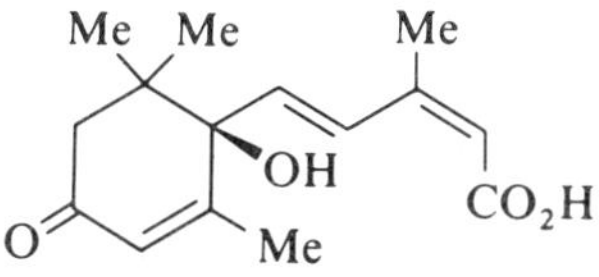

112 Abscisin II

The kinins (cytokinins) are involved in cell division and through it can exert a decisive effect on differentiation and initiation of roots and buds. Kinins abound in fruit and seeds and may be involved in the differentiation of the embryo. They also affect leaf growth, light response, and aging. 6-($\gamma\gamma$-Dimethylallylamino)purine **111** is an example of a natural kinin.

Natural growth inhibitors exist in cotton, sycamore, and yellow lupin. The cotton inhibitor, abscisin II **112**, also occurs in sycamore leaves and in yellow lupin. It induces dormancy in birch leaves and plum twigs and inhibits germination in rose hips.

Apart from IAA **109**, the gibberellins GA_3, GA_4, and GA_7, and ethylene, the commercial growth regulators are synthetic mimics or materials influencing the production or behaviour of the natural hormones. The first important developments were the use of acetylene and ethylene in fruit ripening, the use of IAA and indolebutyric acid in stimulating root production in cuttings, and—most important of all—the use of phenoxyacetics **103** for selective weed-killing in cereals. Plant growth regulators form one of the major growth sectors in agrochemicals and growth regulation is attracting much research effort.

The variety of potential uses for growth regulants discovered, and in many cases exploited, in the 1950s and 60s is astonishing. The benefits may be demonstrated in the management of fruit production. Thus, in apple production 1-naphthalene acetamide (NAD) **113a** and 1-naphthalene acetic acid (NAA) **113b** control preharvest drop by nullifying the effects of imperfect pollination and temporary physiological stress. Gibberellic acid, applied at the appropriate time, promotes lateral branching and rapid establishment of seedling trees, or delays fruit ripening and senescence on older trees. *NN*-Dimethylaminosuccinamic acid, if applied after trees have flowered one year, promotes flowering in the following year. On grapes, GA_3 **110**, applied at the correct stage, regulates the time of harvest and increases the size and yield of seedless grapes. On citrus, the effects of gibberellic acid are just as varied; in low-yielding clementines, yields are greatly increased. A new potential use lies in application to objectionably seedy critrus varieties after destruction of the flower clusters. Delayed ripening, readily achievable with gibberellic acid, has many advantages: it makes harvesting easier, delays senescence, and above all retards the cellular breakdown of the skin that leads to undesirable colour changes and to physiological or pathological disease.

Outlets on vegetable crops include the production of evenly developing Brussels sprouts for the processing industry, assisting in the forcing of rhubarb, controlling the growth of celery to produce an earlier harvest or a later one with enhanced yield.

Some important developments have also occurred on arable crops. In many parts of Europe, a combination of wet summers and high fertility can cause severe lodging of cereals, partly because increased nitrogen uptake gives tall, weak straw. The use of chlormequat **114** can counteract this tendency in wheat. The benefits are less noticeable on other cereals. Cotton, however, is responsive.

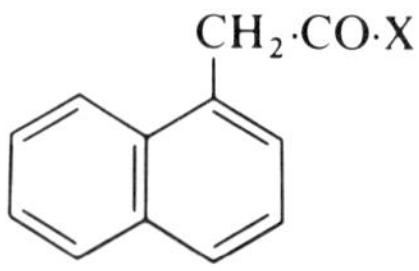

$$\text{naphthyl-CH}_2\cdot\text{CO}\cdot\text{X}$$

113a NAD, X = NH$_2$
   b NAA, X = OH

$$\text{Me}_3\text{N}^+\cdot\text{CH}_2\cdot\text{CH}_2\text{Cl} \quad \text{Cl}^-$$

114 Chlormequat

8. Nutrition of crops   D. PRICE JONES

THE carbon, hydrogen, and oxygen needed for growth are readily available to a plant. In practical crop nutrition, therefore, attention is concentrated on other elements that may be in short supply and thereby affect crop yield or quality. Only thirteen elements are considered essential. Of these, nitrogen, phosphorus, potassium, calcium, magnesium, and sulphur are classified as major essential elements because they are required in relatively large amounts. The remaining seven—iron, boron, manganese, zinc, molybdenum, copper, and chlorine—are required in extremely small amounts and are conveniently called 'trace elements': they may acquire crucial importance on certain soils and under special agronomic conditions.

Of the major elements, nitrogen is present in amino-acids and proteins and many other important substances, including chlorophyll on which all green plants depend for the conversion of solar into chemical energy. Phosphorus enters into nucleoproteins and many compounds involved in metabolic transfer processes; it is particularly important in seedling growth. Potassium is required for the formation of sugars and proteins, but is not incorporated into the end products. It is involved in cell division, in the regulation of nutrient balance, in activating certain enzymes, and in adjusting water relationships. Calcium has a complex role in plant nutrition. It is essential for cell growth and division and commonly influences the activity of other elements. Magnesium enters into the chlorophyll molecule, is involved in the formation of amino-acids and vitamins, fats and sugars, and essential for the germination of seed. Sulphur is incorporated into several amino-acids and vitamins. It is necessary for the synthesis of chlorophyll but is not itself incorporated.

Plants participate in recycling processes, two of which, those relating to carbon and nitrogen, are particularly significant in plant nutrition.

The carbon cycle exchanges carbon between the atmosphere, living organisms, dead organic material and soil and water. Green plants convert carbon dioxide (derived mainly from the atmosphere) into sugars and subsequently into a variety of materials. When the plants, or parts of them, die, microbiological degradation of the material releases carbon dioxide to the atmosphere once more. In an evolutionary sense, the carbon cycle has been essential for ensuring adequate supplies of carbon for plants, and in generating soils in which plants can grow. Historically, soil organic matter has been important, partly because it improved the texture of soils and partly because it helped the plant to make the best use of the available nutrients. These properties are not quite so important in modern agriculture.

Atmospheric nitrogen is not directly available to green plants, except for certain blue–green algae. However, some plants, particularly legumes (peas,

Nutrition of crops 67

beans, clover etc.) have associated bacteria capable of fixing atmospheric nitrogen, which ultimately becomes available to the plant. Some free-living (as opposed to symbiotic) bacteria, *Azotobacter*, can fix nitrogen under aerobic conditions, while *Clostridium* does so anaerobically. Microbiological fixation may account for an annual input of about 20 kg ha^{-1}, while electrical discharges in the atmosphere may account for about 2–3 kg ha^{-1}, in terms of elemental nitrogen. Atmospherically-fixed nitrogen is deposited in the soil by precipitation: it may be in the form of the ammonium ion or oxidized to the nitrate ion. In the soil, nitrifying bacteria promote the oxidation

$$NH_4^+ \dashrightarrow NO_2^- \dashrightarrow NO_3^-$$

Entry into the green plant is mainly as the ions NH_4^+ and NO_3^-. Within the plant, various nitrogen compounds, principally amino-acids and proteins, are synthesized. On the death of the plant, or parts of it, microbiological degradation sets in and amino-acids, including those from protein breakdown, undergo ammonification.

Losses from the soil may be due to *Nitrosomonas* reduction of nitrates to gaseous nitrogen, nitrous oxide, or nitric oxide; or to the liberation of ammonia. Nitrogen is also lost through leaching, soil erosion, and crop removal. Crop removal may result in annual losses of 25 to 50 kg ha^{-1}, more than the average input from natural fixation. Ancient systems of agriculture evolved around the nutrient economy. The river valley systems in Mesopotamia and Egypt depended on river-borne inputs, such as the Nile flood, to make up the deficit. In the tropics, impoverished areas were allowed to recuperate while cultivation was shifted elsewhere. Mediaeval—and subsequent—European systems depended on the husbanding of organic residues with their minerals, concentration of these in favoured areas, and the use of weedy fallows or of leys to exploit natural fixation. By refining these techniques, productivity was pushed to the limits of the nitrogen economy. More intensive development had to await modern manufacture of synthetic fertilizers, particularly of nitrogen.

Today maintenance or improvement of the productivity of the land combines old and new technology. Home-produced farmyard manure still contributes to the conservation of plant nutrient resources. In the 1950s in Britain, 35 to 40 million tons of manure were used annually, equivalent to at least 180 000 tons of N and K_2O and 80 000 tons of P_2O_5. Britain's nutrient resources are also boosted by the purchase of feedstuffs. But in intensive farming it is necessary to maintain the level of nutrients with chemical fertilizers. Fertilizers have assumed overriding importance in recent years as crop production has become more intensive, partly because the handling and transport of bulky organic manure are costly, partly because insufficient manure is available. The amounts of N, K_2O and P_2O_5 applied as chemical fertilizers are about 0·25 million tons per annum each and still rising. These vast amounts of chemical fertilizers are made by a large-scale chemical industry, mostly new or restructured since 1945.

Nitrogenous fertilizers

The nitrogen in commercial fertilizers is derived entirely from the fixation of atmospheric nitrogen. The nitrogen is combined with hydrogen derived from hydrocarbons and steam in the normal operations of the petrochemical industry. Some of the ammonia so made is converted into ammonium salts, some into urea, and some is oxidized to nitric acid. These products are the major ingredients in fertilizers.

Nitrogenous fertilizers include sodium nitrate, ammonium sulphate, ammonium nitrate (with or without the addition of calcium carbonate) and urea and its salts. Although the responses of individual crops differ somewhat, there is little to choose between nitrogen in nitrate or ammonium forms. Urea is highly 'concentrated'—it contains a much higher percentage of N than does ammonium nitrate. It is therefore preferred where transport or application costs are high. Before the nitrogen in urea can be utilized by the plant, conversion into ammonia or nitrate must occur. A high ammonia concentration near the plant can injure it.

Time of application of fertilizers is often crucial. Under British conditions, autumn-applied nitrate is generally lost, either leached or degraded. Ammonia is absorbed into clay particles and is retained if winter temperatures are low; in mild winters, microbiological conversion to leachable nitrate occurs. For spring crops, most nitrogen is therefore applied to the seed-bed with some later applications, particularly to intensively-grown grass. Autumn sown crops receive a seed-bed application with a spring top dressing to follow. Where protracted release from one application is required, so-called slow-acting nitrogenous fertilizers are used.

Phosphate fertilizers

At one time bones and basic slag from steel works were the main sources of phosphatic fertilizers. Bonemeal was used as a 'conditioner' for other fertilizers. Nowadays, rock phosphate—a generic term for a variety of phosphatic ores—is the major source. It is mined mainly in the U.S.A., Siberia, and North Africa. Phosphates are included amongst the world's essential resources: known reserves are not large and may be exhausted before long. Finely ground rock phosphate may be directly applied to moderately or strongly acid soils, particularly for perennial crops. The availability of phosphate in rock phosphate is higher in acid soils with a low calcium content. Rock phosphate is also useful for building up phosphate fertility in soils initially low in phosphate.

Where phosphate is required in a more readily available form, 'superphosphate' is commonly used. This consists of a mixture of monocalcium orthophosphate with some dicalcium phosphate and substantial quantities of calcium sulphate. A large-scale industry produces this mixture from ground rock and strong sulphuric acid. Superphosphate is widely used, particularly on root crops. It has a high content of available phosphorus, much of it water-soluble. 'Triple superphosphate', made by treating rock phosphate with phosphoric acid and thus much more concentrated, is commonly used in compound fertilizers.

Potash fertilizers

Deposits of potassium salts occur in many parts of the world; the main reserves are in the U.S.S.R. and Canada. The U.K. has a potash deposit in North Yorkshire. Previously the Strassfurt deposits in Germany were a main source of potassium salts. Chlorides, particularly sodium chloride, can harm certain crops; hence potassium sulphate is most often used, preferably in a relatively pure form. On sugar beet, however, chloride at normal application rates is not harmful, and potassium chloride is used. Sodium and potassium are largely interchangeable on sugar beet and on certain other crops.

Compound fertilizers

In modern farming, the annual loss of nutrients from the soil demands the frequent use of fertilizers. However, different crops, soils, and husbandry systems require different proportions of nitrogen, phosphorus, and potassium. Though it is impossible to provide a wide enough commercial range of fertilizers to ensure an optimum combination for all conditions, sufficient products are marketed to ensure that at least one is satisfactory for any particular requirement. Compound fertilizers comprise intimate blends of the component materials. Mere dry mixing might result in subsequent segregation in storage, transport, or application. From the blended fertilizer, granules or prills are produced. To increase the concentration and reduce transport costs, the components are selected with care, and those with higher contents, particularly of nitrogen (such as ammonium phosphate and ammonium nitrate) are favoured over mixtures of ammonium sulphate with sodium salts.

Liquid fertilizers

Ammonia is made on a vast scale; world-wide more than 40 million tons are made, worth more than £650 million. Its main use is as an intermediate product in the manufacture of ammonium sulphate or ammonium phosphate. Where it can be used directly, without conversion into a salt, cost-savings accrue. Its effectiveness as a fertilizer has now been widely accepted. However, before it can be transported and applied it has to be liquefied, or dissolved in water. Difficulties in liquefaction, transport, and application have been overcome by the development of suitable equipment. Liquid ammonia has become an important fertilizer in the U.S.A. and, to a much smaller extent, in Europe. The key to the development is the lower winter temperature in those parts where the system is employed. An extended application period is needed to cover the expenditure. Aqueous ammonia, because of its greater bulk, is more expensive to transport, but its handling and application are easier and require less expensive equipment. Where aqueous ammonia is applied, phosphate and potash fertilizers can also be applied in solution. Advances in both techniques may be expected to continue.

9. The industrial development of a new agrochemical D. PRICE JONES

THE production of new agrochemicals nowadays requires large research and development resources. Scientific staff must include chemists, biochemists, biologists, agronomists, toxicologists, pathologists, engineers, mathematicians, and economists. Laboratories have to be fitted with specialized equipment; glasshouses and controlled-climate growth rooms are required. In contrast to pharmaceuticals research, biologists and agronomists working on agrochemicals need land for crop experimentation and facilities for field evaluation on commercial crops in many parts of the world. Ancillary services, including information and computer services are essential.

Pesticides and plant growth regulators

The development pattern is highly variable but that presented in Scheme 2, grossly simplified though it is, serves as a general guide. The stages I–V are arbitrarily delineated to facilitate presentation.

At Stage I the work in progress includes the synthesis of compounds in response to some inspiration—personal or from the literature—or for more mundane reasons, such as the resources available or local know-how. These compounds, in simple formulations, pass through biological screens, acting as general indicators of biological activity. Some specialized screens may also be in operation in the search for solutions to specific problems. Any one screen tends to evolve and to become more effective as the operators gain experience and as deficiencies are rectified.

Sooner or later a new lead is discovered (Stage II). An iterative process then starts: synthesis (to permit structure–activity correlation) according to a definite pattern and submission of the resulting materials, in simple formulations, to the biological screen. Screening now becomes more critical, both in experimental finesse and in the range of dosages employed. Acute toxicity tests towards certain mammals start. The results of the biological and toxicological tests are returned to the common pool and help to guide further synthesis and evaluation. This interaction between synthetic chemists and biologists gradually extends to analytical chemists and biochemists and, at least in consultation, to ecologists and economists.

In due course this interaction leads to the identification of one or a few compounds (Stage III), which are subjected to more thorough physicochemical characterization and formulation. Likely future use has to be assessed: technically this requires definition of physical state, stability under a variety of storage conditions, packaging and suitability for application by the methods envisaged. The biologists evaluate in more detail both the direct biological effects on the target organisms and some of the side-effects, for

SCHEME 2

DISCIPLINE	STAGE I Normal activity†	STAGE II Lead found	STAGE III New material shows promise	STAGE IV Decision to develop	STAGE V Development launch
Chemical synthesis	Many compounds synthesized†	Directed, progressive synthesis		Candidates selected	Patent specifications
Formulation, physical chemistry	Formulation of compounds for screening†	Formulations with more sophistication for wider use	Characterization, formulation	Characterization, formulation	Detailed studies, storage tests
Process chemistry	—	—	Consideration of large-scale processes	Scale-up to provide material; semi-technical or pilot plant	Produce in plant
Biology (laboratory)	Routine screenings†	Wider screening	More detailed evaluation	Comparison, biological studies	All studies
Biology/ agronomy (field)	Study of crops, pests and pest control†	—	Initial appraisal	Further evaluation	

Department				
Analytical chemistry	Quality control; development of new methods†	—	Methods for control and determination of residues/metabolites	Methods, residues	
Biochemistry	Mode of action studies†	—	Initial studies	Metabolism, mode of action	in greater depth, as required
Toxicology	Study of effects and side-effects†	Acute toxicities	Further tests	Toxicological profile	
Ecology	Inter-relation of chemicals, pests and other species†	—	Consultation, initial tests	Laboratory and field tests	
Economics and commercial departments	Assessment of customer needs and value of meeting needs†	—	Consultation on markets	Initial market estimates	
Product development	Improvement of products†	—	—	—	Development starts

† Some of the 'on-going' R & D work at anytime.

instance, the interactions with crop plants or with beneficial organisms, such as parasites, predators or competitors of pests. The biologists also study the mode of action and behaviour of the compounds over a range of simulated climatic conditions. Agronomists assess the performance of the materials in the field and include a comparison with one or more competitive products or alternative techniques for achieving the same end. Persistence, metabolism in plants and animals, and mode of action are examined and analytical methods explored. Toxicology embraces a wider spectrum of tests, mostly of relatively short duration. These tests help to establish a toxicological profile essential in the design of long-term tests and/or further critical investigations. The broader ecological implications are studied: laboratory and field tests tend to concentrate on ecologically significant animals, such as birds, fish, earthworms, and predatory beetles, the implications being pursued wherever they may lead.

Although the final choice of compound may not yet have been made, preliminary economic assessments are needed. First, process chemists establish the feasibility of a commercial process and provide rough production costs at certain levels of production. Agronomists and biologists define the biological and agronomic characteristics so that commercial departments can estimate market potential at given price levels. The economists, already involved in the market survey, then complete the assessment. All this information is required for one crucial decision: to proceed to development, or to abandon. It is never an easy decision—data are incomplete, the financial investment is heavy, and the commitment of resources can cripple other projects. To make matters worse, postponement of a decision increases costs and shortens the effective life of the product.

If the decision is to proceed (Stage IV), the operations are intensified, with major emphasis on the sectors where information is most needed or to solve specific problems. The synthetic chemists may begin to ease out of the development at this stage, or they may be obliged to continue synthesis to provide further material or information for patent purposes. They may go on to experimental production or to process research in preparation for the construction of a pilot plant.

Formulations sufficiently close to the definitive products are needed for field tests, results of which must be suitable for presentation to the appropriate regulating authorities. Biologists still support the project. They may, for instance, examine alternative formulations or, with biochemists, study metabolism, persistence, or mode of action. The weight of evaluation will, however, have shifted to the field, to commercial crops under commercial conditions. This work continues even after the product is launched. In whatever country it is used, it has to be registered and approved. Detailed information has to be submitted relating to the safety and efficiency of the product, much of it obligatorily obtained in the territory concerned, to ensure that local conditions have not been overlooked. Many problems arising in the field demand experimental investigation in the laboratory. Likewise, new discoveries of new ideas within research and development may be transformed into new formulations or new recommendations for the field. The advent of new competitive products can also produce a spate of fresh development effort.

As sales proceed, development work tends to take on a technical service character. The work is increasingly oriented towards serving the customer, usually through the sales organization or through the merchants or agents from whom the customer buys. A great deal of 'trouble-shooting' can be expected in the first years of use of a new product. This is best done initially by those who assisted in the birth and development of the new product. The task falls mainly to biologists and agronomists but formulation chemists are frequently expected to spend a great deal of their time in the field.

It is clear that the development of a new product depends on the interaction of many disciplines—that is, on team work. As a consequence, the management of interdisciplinary projects has become a feature of the agrochemical industry in recent years. There is little published information on the systems adopted in the various firms, but most of them operate project teams, responsibility for which is vested in the project leader, who reports to senior managers and draws resources from wherever required in the enterprise.

Fertilizers

The development of a new fertilizer, has certain similarities to that of a pesticide, but also major differences. The same range of scientific disciplines is involved: also, in broad terms, the same facilities in laboratories, glasshouses, and land. The differences relate to the much more limited scope for novelty in fertilizers and to the relatively direct, uncomplicated manner in which they impinge on the human environment.

The course of development of a new fertilizer depends on the nature of the novelty. If it is a new formulation, routine precautions must be taken to ensure that it is biologically equivalent to the product it is replacing. If specific advantages are claimed (e.g. better storage properties) it will require, e.g. storage under commercial conditions, followed by the necessary tests in the field to ensure that its usefulness is maintained.

Today the availability under field conditions of the plant nutrients in fertilizers is under close scrutiny. In particular, slow release of nitrogen is desirable; hence, much effort has been devoted, for example, to the coating of granules, or the use of urea–formaldehyde formulations which delay degradation and the premature release of nitrogen. Although behaviour of such products in various soils may first be studied in the laboratory, the full implications can be assessed only in the field.

There is one topic where the development of a fertilizer differs in emphasis, if not in principle, from the development of a pesticide. So far, agricultural systems and crop varieties have evolved under conditions of nutrient conservation at relatively low levels. Now that altogether higher levels of nutrients can be maintained, full benefit will be reaped only by breeding new crop varieties capable of responding to the increased input of nutrients. This is now happening throughout the world and is particularly significant in the 'green revolution' in the tropics and subtropics.

10. Agrochemicals and the environment   D. PRICE JONES

No appraisal of the agrochemical industry would be complete without a reference to its interaction with the environment. Because most pests occur in large numbers, control is directed not at individuals but at populations. Thus pesticides are applied to an area or volume, with the object of reaching all, or most, individuals of a particular pest. Chemicals with biological effects are injected into the ecosystem because they cannot be directed specifically at the individual pests. In this respect, pesticides differ fundamentally from medical and veterinary drugs, which are used for individuals and are largely prevented from escaping into the environment. It is almost inevitable that pest control by today's techniques will affect organisms other than the target pest. It is important, therefore, to identify any side-effect that might occur and to take steps to eliminate or at least minimize it.

Historically, the first side-effect considered was toxicity to man and to domestic animals, and also to crop plants. Mammalian toxicity certainly dominated the thinking into the 1940s. In the 1950s, the emphasis swung to consumer hazards arising from pesticide residues in food: first, to cumulative toxicity and later—and even more insidiously—to potential carcinogenicity and perhaps genetic disturbance. In the late 1950s and in the 1960s, observations on the incorporation of pesticides into biological food chains led to a deeper involvement with wildlife and finally with the environment as a whole. Three other problems could not be completely dissociated from the more commonly discussed environmental effects. These were: induction of pesticide resistance in certain pests, resurgence of pests after pesticide usage, and generation of new pests (or elevation of pest status) as a consequence of the use of pesticides.

The toxicity of pesticides has been the subject of some sweeping generalizations. Insecticides seem to be much more toxic than are fungicides and herbicides, although there are notable exceptions. It is important, however, to recognize the hazards associated with particular products rather than with a wide range of products: appropriate precautions can then be taken. Over the last thirty years, as affluence in the developed countries has grown, social structures have developed leading to voluntary or statutory control of new biologically active products. Thus, today, pesticides are marketed only with the approval of the authorities concerned; the recommended application techniques and precautions are agreed beforehand; they are clearly stipulated on labels and in technical literature. With sensible observance of such precautions in Britain there have been no deaths on farms from the use of pesticides in recent years. Most developed countries have similar healthy records. Where highly toxic pesticides are used without proper precautions, fatalities must occur from time to time.

Accidents and misadvantures do still occur: there have been some almost incredible cases of misuse, where toxic materials have been repacked in beer or soft-drink bottles and even stored in domestic refrigerators, with fatal consequences. Under such conditions, there is considerable pressure on the industry to produce materials that, even when deliberately misused, will not prove significantly toxic. Already some inherently toxic materials are highly diluted or made unappetizing. This trend will continue, but a more satisfactory solution is the development of new materials with low toxicity to replace those of higher toxicity. Normal commercial pressures will make this happen, but the actual change may be slow: it depends on the discovery and introduction of new compounds. Innovation in biological effects is essentially slow.

It is recognized that pesticides released into the environment may leave residues. The permitted levels of residues in food are strictly regulated, first by insistence on safeguards at application (dosage, timing in relation to harvest etc.) and secondly by monitoring by public analysts. The movement of foodstuffs across national frontiers, and international co-operation ensure that the most rigorous standards become general.

The effects of a pesticide on wildlife depend mainly on its toxicity and persistence. The public increasingly demands highly selective materials—possibly even species-specific—but the vast numbers of pests make this policy quite unrealistic. Species-specific materials can be envisaged only in connection with a few pests of outstanding importance. In general, a more promising approach is the use of group-selectivity where activity is confined to a number of related species, for example, caterpillars, powdery mildews, and grass weeds. A moderately broad spectrum of activity can be tolerated, provided the material is not too persistent. After the material has performed its function, degradation in living tissues, or in soil or water, is an advantage; so are any properties preventing accumulation in the living organism. Methods of application that help to reduce the impact on wildlife are also desirable. These properties of potential agrochemicals are now examined closely and evaluated by registration authorities in most countries, before approval is granted. Over the last decade, pesticides have become much safer in use.

Yet this piecemeal reduction of harmful side-effects of pesticides constitutes a rather negative approach to pest control and crop protection. Many ecologists prefer a different approach; they consider the whole ecosystem and think of pest control as some aspect of its regulation. An organism (even a noxious organism) is in balance with its environment and that balance can generally be changed in man's favour, with minimal damage, by manipulating the environment. Applied to practice, these ideas are achieving some success.

Agrochemicals and the environment

The systems ecologist regards pesticides merely as interim devices in lieu of better alternatives. He approaches pest control through the population dynamics of the pest and of parasites and predators with which the pest interacts strongly. He attempts to model the system concerned and to identify the major parameters. With the aid of such a model, he attempts to predict effects of manipulating the ecosystem and so deduces promising methods for regulating the pest. It is no serious criticism that the model is only a rough approximation; if it leads through field evaluation of particular methods to the selection of an effective technique, it will have served its immediate purpose. Refinements will afford improvements later. At its present stage of development, the systems approach is most valuable as a guide for those biologists primarily engaged in a more empirical field approach.

Another approach, historically different from the systems approach, but moving towards a similar solution, is that known as 'integrated control'. Originally this consisted of manipulating the pesticide input to make full use of the natural biological control agents (mainly parasites and predators). In this sense it was strictly the integration of chemical with biological control. This restricted integrated control has been successful with several crops, especially alfalfa and cotton. Limited success has also been achieved in some fruit crops, where the pest problems are acute and complex.

The concept of integrated control has been extended to any combination of control measures against one or more pests on a given crop. Where true integration was envisaged, it necessarily involved thinking in terms of systems. In practice true integration has rarely been achieved over a broad array of control measures. More typically, all available control measures have been applied with little genuine integration except perhaps between biological control on the one hand and pesticides or husbandry practices on the other. Thus aphid-borne viruses on sugar beet have been controlled by a variety of measures (e.g. control of overwintering hosts of viruses and aphids, isolation of seed crops from sugar production areas, monitoring of aphid and virus abundance and controlled applications of aphicides) without deliberate integration.

It is now widely recognized that the 'balance of nature' has to be considered whatever method of pest control is envisaged. Thus, the eradication of one species of malaria-transmitting mosquito from Sicily some years ago merely resulted in another species occupying the vacant niche. In the island of Santa Cruz, overgrazing of pastures by sheep caused a great upsurge in the abundance of prickly pear which was ultimately controlled by the introduction of the cochineal insect. This enabled the grass to recover and then assisted in suppressing the prickly pear. In the past, broad-spectrum insecticides used against certain major pests, especially on cotton, have also killed beneficial insects that helped to suppress other, initially minor, pests, which then increased greatly in importance. The effects of interference with an ecosystem are now studied as a matter of course. As results are interpreted, the agrochemicals industry is learning where to look for these side-effects, how to minimize them and what early warning systems to establish.

Fertilizers also interact with the environment but in a manner quite different from pesticides. The nutrients supplied by fertilizers are precisely those around which plant biochemistry has evolved over hundreds of millions of years and as such they are inextricably woven into the web of life. Any adverse effects of fertilizer usage should therefore be sought in the changing balance of plant growth and ecosystem productivity. As long as applications are properly controlled (and the tolerance is high), crop production is increased without significant harm to man or beast. Only in rare cases are there dangers to man. For example, spinach over-generously fertilized with nitrogen, and harvested early, contains enough nitrate to be hazardous to young babies: the acidity of the infant stomach is insufficient to prevent microbial reduction of nitrate to nitrite, which combines with haemoglobin in the blood stream to produce the methaemoglobaemia of 'blue babies'.

Where fertilizer usage is heavy so that some escapes into adjacent waterways, the enrichment of the water (referred to as 'eutrophication') can lead to blue-green algal blooms and subsequent harm to aquatic life. In Britain this is very rare and limited to occasional escapes of effluent from intensive animal-husbandry systems or from town sewage works. Water authorities maintain a close watch on the nitrate and phosphate content of their water supplies.

The layman tends to confuse fertilizer use with 'intensive farming' and to attribute some of the known ecological effects of such farming to fertilizers, it is true that fertilizers are an essential part of most systems of intensive farming but, apart from this, they are hardly implicated in the impoverishment of the environment that sometimes accompanies intensive agriculture.

11. The future of the industry
D. PRICE JONES

THE food supply for the poorer nations is inadequate. It would become disastrously so without the use of fertilizers to increase production, and of pesticides to safeguard it. Without pesticides, insect-borne diseases would again devastate human populations in the tropics. The agrochemical industry thus has an assured future. Yet the industry is young; changes in both pesticide and fertilizer practice will continue. Where will these changes lead?

Some feel that the changes, now that the agrochemical industry may be regarded as a distinct entity, will be concerned mainly with its interaction with the community. As the industry has contributed so significantly to the community, so the community has inevitably—and justifiably—increased its control over the industry. Industries affecting health and food are proper subjects for governmental concern. Yet the agrochemical industry is international and its benefits and side-effects do not stop at national boundaries. These are two conflicting, but not necessarily irreconcilable aspects of the industry. A degree of easement is provided, for example, by the growing tendency of certain governments to adopt similar standards, but this process still has a long way to go. In the long term, the agrochemical industry must develop an ever closer relationship with the community, its plans must be more closely integrated with the policies of international and supranational organizations, and some of its functions (especially some of its research and development) must increasingly meet the objectives of national and international agencies, while continuing to meet the needs of individuals and of agriculture.

Pesticides

Even in the past ten or fifteen years the philosophy of pest control has changed greatly. The emphasis has switched from killing pests to managing pest populations, with due regard to human and environmental needs. This trend will continue for some time, with perhaps some abrupt checks during periods of near-famine, and in deprived areas where immediate food and health problems predominate. Tangible expressions of this evolving philosophy are discussed in the preceding sections on the systems approach to ecology and on attempts at integrated control of pests. They pinpoint much fruitful work for chemists and biologists. Synthetic chemists will search for new types of biological activity in novel chemical systems, using the ever increasing knowledge of activity in relation to molecular structure. Formulation control has already emerged from the development stage familiarly known as 'cookery' and must become even more precise. With the increasing public concern over trace-quantities of noxious materials in the environment, analytical chemists have the important tasks of developing and standardizing methods for monitoring chemical residues. Similarly, increasing demands for information on pesticide metabolism and mode of action will create more opportunities for biochemists. It should not, however, be assumed that the pesticide industry's involvement with pest control in the future will be restricted to pesticides as such; chemicals which are not lethal agents are already in use. So-called 'repellents' have been marketed commercially for many years. Recently, both these and attractants (for use with poisons) have been much studied. Other chemicals with subtle influence on pest behaviour are beginning to receive attention. These include, for instance, the sex pheromones that play such an important part in the mating of insects. Chemicals with sterilizing properties are being used—tentatively, as yet—for the control or elimination of pest populations. Also, the part of the chemical industry with fermentation interests (mainly production of antibiotics) is already marketing microbial insecticides, thus partly bridging the gap between purely chemical control and purely biological control.

The production of basic chemicals for the pesticide industry is still largely concentrated in developed countries. However, formulating plants have to be close to the market they serve. This is likely to lead to some increase in production of basic materials in developing countries, which are heavy users of insecticides, partly because of the importance of insects as carriers of human disease. Developing countries generally have a surplus of labour and therefore little incentive to develop the use of weed-killers. It may be some time before the overall balance changes, but some local changes have already occurred.

Plant growth regulants

The scope for regulating the growth of crop plants in the interests of the grower and the community is immense. The variety of effects almost defies description. Many of these have been demonstrated only experimentally and commercial use may be a long way off. Nevertheless, while the farmer continues at the mercy of the weather, he will eagerly grasp any help he can get from plant growth regulants, as long as they represent good value for money. Similarly the economic pressures on farming favour techniques, including the application of chemicals, that increase yield, improve quality, ease harvesting, or assist marketing. There are plenty of targets for the economically conscious chemist here.

Fertilizers

The fertilizer industry, too, is expecting continued growth. The increasing world demand for meat will encourage more intensive grass management and greatly increased use of fertilizers, especially nitrogen. Secondly, cereals for human consumption and for meat production will be increasingly dominated

The future of the industry

by the new dwarf varieties able to respond to heavy fertilizer applications with minimal risk of 'lodging' (falling over in storms). This latter change is likely to be more dramatic in the less developed countries where fertilizer usage has hitherto been inhibited by economic constraints and traditional practices, as well as by the traditional varieties ill-adapted to enriched soils.

Although the fertilizer industry operates in a world-wide market, production capacity is heavily concentrated in the developed countries. The problem of long distance transport has been eased slightly by the use of 'concentrated' forms of nitrogen (ammonium nitrate and urea) but the needs of the developing countries are such that production must be developed locally in parts of the world where it does not exist today. Ammonia plants are likely to be built near or on oil fields, to use refinery gas or natural gas, otherwise wasted. This inevitably means exploiting also the local resources of phosphates and potash. For the past six decades, industrial nitrogen has been produced by the Haver-Bosch process, which uses high pressures and a considerable energy input. Research on new systems of nitrogen fixation by metal complexes based on natural models, is in progress. If successful, such a system would have the commercial advantage over the Haber-Bosch process, of requiring a lower consumption of power and less complicated plant (and thus less initial capital): this would help the poorer countries.

The two major ecological constraints on the increased use of fertilizers are the dangers of eutrophication and of highly localized cases of nitrite poisoning due to contamination of drinking water or to gross over-fertilization of certain vegetable crops used for baby food. Both these constraints are amenable to sensible control.

Further reading

Pharmaceuticals

SCHUELER, F. W. (ed). *Molecular modification in drug design.* Proceedings of a symposium by the Division of Medicinal Chemistry, at the 145th meeting of the American Chemical Society, Washington, 1964 (Advances in chemistry series, No. 45).

The science of drug discovery, and drug discovery and development in a changing society. Proceedings of symposia by the Division of Medicinal Chemistry, at the 206th meeting of the American Chemical Society, Washington, 1971. (Advances in chemistry series, No. 108).

WATSON, J. D. *The double helix*—a personal account of the discovery of the structure of DNA, Weidenfeld and Nicolson, London, 1968.

GILBERT, J. N. T. and SHARP, L. K. *Pharmaceuticals,* Butterworths, London, 1971 (Chemistry in Modern Industry Series).

Focus on pharmaceuticals. National Economic Development Office, HMSO, London, 1972.

TEELING-SMITH, G. (ed). *Economics and innovation in the pharmaceutical industry,* Office of Health Education, London, 1972.

Gaps in technology—pharmaceuticals. OECD, Paris, 1969.

The pesticide industry

HARTLEY, G. S. and WEST, T. F. *Chemicals for pest control.* Pergamon, 1969.
HASSALL, K. A. *World crop protection.* Vol. 2 Pesticides. Iliffe, 1969.
MELLANBY, K. *Pesticides and pollution.* Collins, 1970.
MARTIN, H. *The scientific principles of crop protection.* Arnold, 6th edn 1973.
National Academy of Sciences. *Pest control: strategies for the future.* N.A.S., 1972.
WHITE-STEVENS, R. *Pesticides in the environment.* Vol. 1 Pts. I & II. Dekker, 1971.

Index

abscisins, 63
acetazolamide, 25
addiction, 11
   liability, 12
Addison's disease, 28
adrenal hormones, 28
adrenaline, 15
aflotoxin, 56
aldrin, 52
7-aminocephalosporanic acid, 22
6-aminopenicillanic acid, 20
aminopterin, 17
ampicillin, 20
analgesics, 10, 11
antagonists
   to folic acid, 17
   to histamine, 15
   to morphine, 13
antibiotics, 19
anticholinergics, 10
anticoagulants, 55
antidepressants, 33
antihistamines, 15, 33
antihypertensives, 16, 34, 37
antimalarial, 2, 9, 24
antimetabolites, 17
6-APA, 20
aphids, 53
apple coddling moth, 51, 52
arsenicals, 52, 55
arthritis, 28
atropine, 3, 10
auxins, 61, 62

Bakanae disease, 63
barban, 61
barbiturates, 38
bendroflumethiazide, 25
benomil, 59
benzodiazepines, 10, 31
benzomorphan, 12
benzothiadiazines, 25
betamethasone, 28, 29
bethanidine, 35
binacapryl, 58
bipyridilium herbicides, 61
blood sugar, 26
blood pressure, 15, 34, 39

bretylium, 35
bunt disease, 56

cancer, 44
captan, 59
carbamates, 50, 52, 54
carbaryl, 54
carbenicillin, 21
carbonic anhydrase, 24, 25
carboxin, 59
carbutamide, 26
cell elongation, 63
chloranil, 58
chlordiazepoxide, 10, 31
chlormequat, 65
chlorpromazine, 11, 15, 33, 37
chlorpropamine, 26
chlorthiazide, 25
choriocarcinoma, 17
cinchona, 2, 9
circulatory disorders, 34
citric acid, 7
clinical trial, 39
cloxacillin, 20
commensals, 46, 55
conception, 29
contraception, 29, 30, 37
copper sulphate
   as fungicide, 56
corticosteroids, 28
cortisone, 28
cotton blight, 56
clorare, 34

dapsone, 24
DDT, 52
decamethonium, 34
demelon, 53
depression, 34
derris, 51
diabetes, 26, 27
diazepam, 31, 32
diazines, 61
dichlone, 58
dinitrophenol, 58, 61
dinocap, 58
diquat, 61
dithiocarbamate, 57

Index

diuretics, 24
dopa, 15, 16
dopamine, 15, 16
drug design, 8
 resistance, 22

endogenous depression, 34
environment, 77
enzymes, 15, 24, 27, 48
epilepsy, 10
ethicals, 2
ethinyloestradiol, 29, 30
ethirimol, 59

ferbam, 57
fertility, 29
fertilizers, 68
fluoracetamide, 55
folic acid, 17
food chains, 52, 57, 77
formulation, 39, 72, 74
Free-Wilson theory, 18
fruit production, 64
frusemide, 26
fungicides, 56

ganglionic block, 34
genetic damage, 38
geriatric disease, 42
gibberellins, 63
glibenclamide, 27
griseofulvin, 40, 59
growth regulants, 48
guanethidine, 35

habituation, 11
hallucinogen, nalorphine as, 13
heart disease, 36, 44
herbicides, 60
heroin, 11, 12
hexamethonium, 34
hibernation, 33
histamine, 14
hormones, 27
 adrenal, 28
 corticosteroid, 28
 pancreatic, 27
 sex, 29
 weed killing, 61
hypersensitivity, 14
hypertension, 34
hypnotic action, 32, 37
hypoglycaemics, 26, 27, 37

imipramine, 33
indoleacetic acid, 62

insecticides, 47, 50ff, 58
insulin, 26, 27

kinins, 63

late blight, 56, 58
lead arsenate, 51
leaf rust, 56
leprosy, 24
leukaemia, 17
louse, 52
lorazepan, 32

malaria, 2, 10, 52
malathion, 53
M & B 693, 23
mecylamine, 35
menazon, 53
meprobamate, 31
mercurials, 24, 57
methicillin, 20
methotrexate, 17
methyldopa, 15
mildew, 56, 58
molecular-orbital approach, 18
morphinan, 12
morphine, 9, 11
mosquito, 52
muscle paralysant, 34

nalorphine, 12, 13
naphthalene acetamide, 64
naphthalene acetic acid, 64
nerve impulses, 15, 34, 54
neuromuscular block, 34
nicotine, 51
nitrazepam, 32
nitrite poisoning, 80, 83
nitrogen fixation, 67
noradrenaline, 15
norgestrel, 29

oestradiol, 29
oral contraception, 29
organochlorine insecticides, 52
organophosphates, 50, 52, 53
oxazepam, 32
oxyphenbutazone, 39
oxytocin, 40

PABA, 17
pain, 10
pancreas, 26, 27
paracetamol, 11
paraquat, 62

parathion, 53
pempidine, 35
penicillin, 7, 19
penicillinase, 20
pentazocine, 13
pesticides, 47
pethidine, 10
phenazocine, 12
pheneticillin, 26
phenobarbitone, 10, 11
phenothiazines, 33
phenoxyacetic acids, 49, 61
phenoxybutyrics, 61
phenoxypropionics, 61
phenylalanine, 15
phenylbutazone, 10, 39
phosphorus as rodenticide, 55
phototropism, 63
phthalylsulphathiazole, 23
phytotoxicity, 48, 56, 57, 58
pirimicarb, 54
placebo effect, 39
plant growth, 48, 62
pneumonia, 2
population control, 30, 79
progesterone, 29
progestin, 29
prontosil, 23
propham, 61
proprietaries, 3
prostaglandin, 30
psoriasis, 17
psychotic disorders, 33
purity, 39
pyrethrum, 2, 51
pyrimidines, 50

quinine, 9

red squill, 55
reserpine, 9
resistance,
  to drugs, 22
  to fungicides, 60
  to pesticides, 77
respiratory depression, 11, 13
rheumatism, 10, 44
rodenticides, 55
rotenone, 50

safety, 37, 43
schizophrenia, 10
  in mice, 37, 47
schradan, 53

scilliroside, 55
screening, 36, 71ff.
sedative action, 32
selectivity, 2, 8, 17, 19, 29, 32, 34, 54, 61
serendipity, 8
side effects, 3, 8, 12, 18, 27, 29, 35, 37, 38,
    47, 52, 77
  of morphine, 11
  of nalorphine, 13
  of organochlorines, 52
  of pesticides, 52
  of sex hormones, 29
stability, 40
streptomycin, 7
structure–activity, 25
strychnine, 55
sulphadiazine, 23
sulphamethoxazole, 24
sulphanilamide, 23, 24
sulphapyridine, 23
sulphathiazole, 23
sulphisoxazole, 23
sulphonamides, 7, 17, 22ff.
sympathomimetic amines, 15
systemic action, 53, 59

TEPP, 53
teratology, 38
tetracyclin, 7
thalidomide, 37
thallium sulphate, 55
thiocarbamates, 61
thiophanate, 59
tolbutamide, 26
tranquillizers, 10, 15, 33
  minor, 31
triazines, 50, 61
tridimorph, 59
trimethoprin, 24
tubocurarine, 34
typhus, 52
tyrosine, 16

urea, 49
  substituted, 50

vitamin K_1, 55

warfarin, 55
weedkillers, 58, 60

zinc phosphide, 55
zineb, 58
ziram, 57

COLLEGE OF MARIN LIBRARY
3 2555 00046444 1

V   40884

TP
1
S7
v.2

Date Due

The Library

COLLEGE of MARIN

Kentfield, California
PRINTED IN U.S.A.

COM